Bruna Missagia

Agricultural and forestry residues in Brazil

Bruna Missagia

Agricultural and forestry residues in Brazil

Decentralized energy generation

Südwestdeutscher Verlag für Hochschulschriften

Impressum/Imprint (nur für Deutschland/only for Germany)
Bibliografische Information der Deutschen Nationalbibliothek: Die Deutsche Nationalbibliothek verzeichnet diese Publikation in der Deutschen Nationalbibliografie; detaillierte bibliografische Daten sind im Internet über http://dnb.d-nb.de abrufbar.
Alle in diesem Buch genannten Marken und Produktnamen unterliegen warenzeichen-, marken- oder patentrechtlichem Schutz bzw. sind Warenzeichen oder eingetragene Warenzeichen der jeweiligen Inhaber. Die Wiedergabe von Marken, Produktnamen, Gebrauchsnamen, Handelsnamen, Warenbezeichnungen u.s.w. in diesem Werk berechtigt auch ohne besondere Kennzeichnung nicht zu der Annahme, dass solche Namen im Sinne der Warenzeichen- und Markenschutzgesetzgebung als frei zu betrachten wären und daher von jedermann benutzt werden dürften.

Coverbild: www.ingimage.com

Verlag: Südwestdeutscher Verlag für Hochschulschriften GmbH & Co. KG
Heinrich-Böcking-Str. 6-8, 66121 Saarbrücken, Deutschland
Telefon +49 681 37 20 271-1, Telefax +49 681 37 20 271-0
Email: info@svh-verlag.de

Approved by: Cottbus, TU, Diss., 2012

Herstellung in Deutschland (siehe letzte Seite)
ISBN: 978-3-8381-3465-9

Imprint (only for USA, GB)
Bibliographic information published by the Deutsche Nationalbibliothek: The Deutsche Nationalbibliothek lists this publication in the Deutsche Nationalbibliografie; detailed bibliographic data are available in the Internet at http://dnb.d-nb.de.
Any brand names and product names mentioned in this book are subject to trademark, brand or patent protection and are trademarks or registered trademarks of their respective holders. The use of brand names, product names, common names, trade names, product descriptions etc. even without a particular marking in this works is in no way to be construed to mean that such names may be regarded as unrestricted in respect of trademark and brand protection legislation and could thus be used by anyone.

Cover image: www.ingimage.com

Publisher: Südwestdeutscher Verlag für Hochschulschriften GmbH & Co. KG
Heinrich-Böcking-Str. 6-8, 66121 Saarbrücken, Germany
Phone +49 681 37 20 271-1, Fax +49 681 37 20 271-0
Email: info@svh-verlag.de

Printed in the U.S.A.
Printed in the U.K. by (see last page)
ISBN: 978-3-8381-3465-9

Copyright © 2012 by the author and Südwestdeutscher Verlag für Hochschulschriften GmbH & Co. KG and licensors
All rights reserved. Saarbrücken 2012

Table of Contents

Table of Contents 1
Terms and abbreviations 4
List of figures 5
List of tables 7
1 Introduction 9
    1.1 Developments in biomass research 11
    1.2 Relevance 13
    1.3 Objectives 15
    1.4 Methodology of work 16
    1.5 Structure of the thesis 17
    1.6 References 18
2 Biomass and energy generation in Brazil 25
    2.1 Historical background 26
    2.2 The potential of biomass in Brazil 28
        2.2.1 The coffee tradition in Brazil 29
        2.2.2 Sugar cane and ethanol 30
        2.2.3 The rice industry in South Brazil 35
        2.2.4 Forestry 36
            2.2.4.1 Native forests 37
            2.2.4.2 Short-rotation planted forests 39
    2.3 The Brazilian energy context 41
        2.3.1 The power sector 43
        2.3.2 Electricity distribution and transmission in Brazil 46
        2.3.3 The Brazilian grid systems 46
        2.3.4 Electricity demand and supply 48
        2.3.5 Electrification programs 49
        2.3.6 Policies and measures 51
        2.3.7 Auction schemes for power generation projects 53
    2.4 Discussion 54
    2.5 References 56
3 Energy from Biomass 66
    3.1 Global energy production and climate change 68

- 3.2 Political framework in Europe 70
- 3.3 Advantages of solid biomass 71
- 3.4 Biomass for heat and power 72
  - 3.4.1 Energy crops 74
  - 3.4.2 Agricultural and forestry residues 74
  - 3.4.3 Biomass logistics 75
  - 3.4.4 Biomass compaction 78
    - 3.4.4.1 Drying 80
    - 3.4.4.2 Grinding 81
    - 3.4.4.3 Moistening 82
    - 3.4.4.4 Briquetting and pelleting 83
  - 3.4.5 Biomass power plant technologies 84
  - 3.4.6 Combined heat and power generation (CHP) 90
  - 3.4.7 Pellet production using CHP 91
- 3.5 Discussion 93
- 3.6 References 94

4 Properties of Brazilian biomass residues 102
- 4.1 Elementary analysis of biomass residues 103
  - 4.1.1 Material and methods 105
  - 4.1.2 Results 105
- 4.2 Agglomeration behaviour 110
  - 4.2.1 Material and methods 111
  - 4.2.2 Results 113
    - 4.2.2.1 Particle size distribution analysis 114
    - 4.2.2.2 Pellet stability: visually noticeable 116
    - 4.2.2.3 Pellet abrasion 117
- 4.3 Discussion 122
- 4.4 References 123

5 Economic feasibility of biomass use for energy generation in Brazil 128
- 5.1 Economic feasibility: theoretical background 129
  - 5.1.1 Methods for investment appraisals 129
- 5.2 General approach for economical feasibility of CHP plants 131
  - 5.2.1 Biomass supply and process chain 135
    - 5.2.1.1 Biomass supply: coffee husks 137

  5.2.1.2 Biomass supply: rice husks 139

  5.2.1.3 Biomass supply: eucalyptus sawdust 139

  5.2.1.4 Biomass supply: Amazonas sawdust 139

  5.2.1.5 Biomass supply: sugar cane bagasse 140

 5.3 Biomass utilization alternatives for Brazil 142

  5.3.1 Compaction case for biomass residues utilization 144

   5.3.1.1 Methodology for compaction case calculations 145

   5.3.1.2 Results 149

  5.3.2 Power generation case for biomass residues utilization 152

   5.3.2.1 Methodology for power generation case calculations 153

   5.3.2.2 Results 157

  5.3.3 Integrated case for biomass residues utilization 163

 5.4 Discussion 168

 5.5 References 170

6 Decentralized biofuel schemes in Brazil 175

 6.1 Approaching smallholders' views and concerns 176

  6.1.1 Biomass utilization schemes in Minas Gerais 179

  6.1.2 Biomass utilization schemes in Pará 183

 6.2 Practical application of results into further projects 190

  6.2.1 Technology 191

  6.2.2 Socio-economic conditions 192

  6.2.3 Environment 193

  6.2.4 Policies 196

  6.2.5 Financing: R&D Projects 197

  6.2.6 Participation of stakeholders 199

 6.3 Discussion 200

 6.4 References 201

7 Conclusions 207

8 Annexes 212

Terms and abbreviations

ANEEL - Brazilian Electricty Regulatory Agency

BMU - German Federal Ministry for the Environment, Nature Conservation and Nuclear Safety

BNDES - Brazilian National Bank of Development

BRL - Brazilian currency (Brazilian real): 1 EURO = 2.26 BRL

CELESC - Electricity Provider of Santa Catarina

CELPA - Electricity Provider of Pará

CEMIG - Electricity Provider of Minas Gerias

DIN - German Institute for Standardisation

EEG - German Renewable Energy Sources Act

EIA - U.S. Energy Information Administration

ELETROBRÁS - Brazilian Central Energy Agency

EMBRAPA - Brazilian Agriculture Agency

EPE - Brazilian Electricty Research Agency

EU - European Union

FAO - Food and Agriculture Organization

FINEP - Brazilian Financing for Research and Projects

FNR - German Agency for Renewable Sources of Raw Materials

GEMIS - Global Emission Model for Integrated Systems

IBGE - Brazilian Institute of Geography and Statistics

IEA - International Energy Agency

IPCC - Intergovernmental Panel on Climate Change

INPA - Instituto Nacional de Pesquisas da Amazônia

MAPA - Brazlilian Ministry of Agriculture, Livestock and Food Supply

MME - Brazilian Ministry of Mines and Energy

OECD - Organization for Economic Cooperation and Development

PROINFA - Program of Incentives for Alternative Electricity Sources

UFPA - Federal University of Pará

UNEP - United Nations Environment Program

UNICA - Brazilian Sugar cane Industry Association

UNFCCC / CDM - United Nations Framework Convention on Climate Change - Clean Development Mechanisms

List of figures

Figure 2.1: Sugar cane production by country (FAO in WADE, 2004) 30
Figure 2.2: Sugar cane: source of energy (Source: UNICA in Jank, 2008) 32
Figure 2.3: Average precipitation levels in Brazil (CPTEC, 2011) 33
Figure 2.4: Wood residues available (m^3 / year) in Pará (PA) (Rendeiro, 2010) 37
Figure 2.5: Wood residues disposed in a football field in Pará (own source) 38
Figure 2.6: Small-scale use of biomass: firewood stove (Fogão a Lenha: Fotos e Modelos, 2010) 40
Figure 2.7: Internal energy supply in 2009 (EPE, 2010) 41
Figure 2.8: Wood consumption in Brazil (MME, 2007) 42
Figure 2.9: Domestic electricity supply in 2009 (EPE, 2010) 44
Figure 2.10: Biomass power plants installed capacity in each state and main crop locations (ANEEL, 2008) 45
Figure 2.11: The National Grid (ANEEL, 2008) 47
Figure 2.12: Production costs of electricity in Brazil based on different raw materials (ANEEL, 2008) 48
Figure 3.1: Fuel shares of total primary energy supply in 2008 (Source: IEA, 2010) 68
Figure 3.2: Evolution from 1971 to 2008 of world total primary energy supply by fuel (Mtoe) (IEA, 2010) 68
Figure 3.3: Global trends in carbon dioxide emissions from fuel combustion by region from 1971 to 2004 (Source: IPCC, 2007 a) 69
Figure 3.4: Mechanization of short rotation woody crops (adapted from Kaltschmitt et al., 2009) 76
Figure 3.5: Black wood chip on the left and white wood chip on the right (own source) 76
Figure 3.6: Herbaceous combustible material allocation (adapted from Van Loo and Koppejan, 2008) 77
Figure 3.7: a) At low pressure, b) At high pressure with brittle material, c) At high pressure with elastic and plastic material (adapted from Pietsch 1991) 78
Figure 3.8: Straw pellets (own source) 79
Figure 3.9: Briquetting devices (hydraulic on the right and mechanic on the left) (Kaltschmitt et al., 2009) 83
Figure 3.10: Pelleting device (Kaltschmitt, 2009) 84
Figure 3.11: Reciprocating grate furnace (McGowan, 2009) 88

Figure 3.12: Package fluidized bed boiler (McGowan, 2009) 89
Figure 3.13: Overall efficiency of conventional CHP power plant (adapted from Boyle 2003) 91
Figure 4.1: (A) Moisture content after 6 week drying; (B) Ash content (water free); (C) Lower calorific value and (D) Ash melting point (Missagia et al. 2011) 106
Figure 4.2: Guiding values for unproblematic thermal utilization (DIN, 2010) and sulphur concentration in biomass fuels 109
Figure 4.3: Guiding values for unproblematic thermal utilization (DIN, 2010) and chlorine concentrations in biomass fuels 110
Figure 4.4: Raw biomass: rice husks (A); coffee husks (B); Eucalyptus saw wood (C); sugar cane bagasse (D) 111
Figure 4.5: Grinded biomass (4 mm): sugar cane bagasse (A) and rice husks (B) 111
Figure 4.6: Laboratory pelleting device. Source: BEPEX - L 200 / 50 G + K 113
Figure 4.7: Extrusion of biomass residues into pellets. Source: BEPEX - L 200 / 50 G + K 113
Figure 4.8: Particle size distribution curve for coffee husks and saw wood (raw form). 114
Figure 4.9: Particle size distribution for rice husks and sugar cane bagasse (4mm grinding) 115
Figure 4.10: Pellets from Eucalyptus saw wood 115
Figure 4.11: Pellets from rice husks, applying less force (A) and more force (B) 117
Figure 4.12: Pellets from sugar cane bagasse grinded to 2 mm (A) and 4 mm (B) 117
Figure 4.13: Compression stress principle 119
Figure 4.14: Three point pressure (left) and point pressure (right) principle 119
Figure 4.15: Results of the abrasion test of rice husks pellets (abrasion percentage) 120
Figure 4.16: Results of the compression stress tests of rice husks pellets (load in N) 120
Figure 4.17: Results of the point pressure tests of rice husks pellets (load in N) 121
Figure 4.18: Results of the three point pressure tests of rice husks pellets (load in N) .. 121
Figure 5.1: Annual load curve scheme for CHP facilities (adapted from Brauer, 2009) 132
Figure 5.2: CO_2 emissions from biomass electricity compared to electricity mixtures (adapted from EPE, 2010) 134
Figure 5.3: Dried coffee (left) and coffee husks (right). (own source) 137
Figure 5.4: Coffee husk furnace (on the left) for drying coffee beans (on the right) 138
Figure 5.5: Sawmill in Espírito Santo (own source) 139
Figure 5.6: Wood log transportation in Amazonas 140

Figure 5.7: Sugar cane harvest (on the left) and milling (own source) 140
Figure 5.8: Sugar cane bagasse (on the left) and bagasse combustion for cooking sugar cane juice (own source) 141
Figure 5.9: Analysis approach for agriculture and forestry residues and production residues (own source) 143
Figure 5.10: Selected scenarios for biomass utilization in Brazil (own source) 144
Figure 5.11: Comparison of production costs for different pelleting approaches. 150
Figure 5.12: Comparison of production costs for different briquetting approaches. 151
Figure 5.13: Example payback calculation of an ORC power plant (200 kW$_{el}$) 154
Figure 5.14: Economic feasibility for biomass plants using sugar cane bagasse (w 50 %) (left) and coffee husks (right), considering the self-financed approach 158
Figure 5.15: Economic feasibility for sugar cane bagasse plant, considering grand fathering approach and external electricity usage (left), internal electricity usage (right) 160
Figure 5.16: Electricity generation costs for 200 kW$_{el}$ nominal power plant under Brazilian price conditions: CEMIG (black line), CELPA (grey line) and auction scheme (dashed line) and different scenarios: (a) DST and self-financed; (b) DST and grand fathering; (c) ORC and self-financed and (d) ORC and grand fathering 161
Figure 6.1: Formal sugar storage (left) and new storage and administration building (right) 181
Figure 6.2: Formal pan for juice cooking (left) and new "double pan" (right). 182
Figure 6.3: Boat transportation (left). Ice and fish handling (right) 185
Figure 6.4: Typical house located in the Marajó Island (left) and daily life (right)...... 185
Figure 6.5: Integrated biomass utilization scheme: sawmill (left) and sawdust CHP plant (right). 186
Figure 6.6: Integrated biomass utilization scheme: oil extraction (left) and muru-murú seed (right). 186

List of tables

Table 2.1: Main sources of agriculture residues in Brazil (Couto el. al., 2004) 28
Table 2.2: Sugar cane planted area: Brazil and São Paulo (IBGE, 2011) 34
Table 2.3: Electric energy consumption supplied by the grid (GWh) (EPE, 2011) 42
Table 3.1: Biomass power technologies in commercial/demonstration phase 84

Table 4.1: Physical-mechanical characteristics of biofuels and their relevance (adapted from Hartmann 2004) 104

Table 4.2: Ash elementary analysis of Brazilian biomass 108

Table 5.1: Technical and economical variables for chosen Brazilian solid biomass fuels 133

Table 5.2: Possible conversion technologies for biomass combustion (Gaderer, 2009 and Schwarz, 2009); acreage calculated according to Kaltschmitt (2009) 134

Table 5.3: Investigated biomass supply chains in Brazil and biomass properties 141

Table 5.4: Input data for cost comparison method 146

Table 5.5: Description of compaction machineries for further cost comparison analysis 147

Table 5.6 Conversion technologies for biomass combustion (adapted from Gaderer, 2009 and Schwarz, 2009) 152

Table 5.7: Cost factors for calculating the electricity generation case 155

Table 5.8: Economic feasibility for different financing approaches, technologies and biomass 162

Table 5.9: Economic analysis of a 7 kW_{el} sugar cane bagasse CHP plant with Stirling technology 165

Table 5.10: Economic analysis of a self-financed 200 kW_{el} CHP plant with ORC technology using native wood sawdust in Pará 167

Table 6.1: Screening of five biomass utilization schemes in six different case studies in Brazil (own source) 177

Table 6.2: Long-term impacts of five biomass schemes in Brazil (own source) 178

Table 6.3 Decentralized biomass generation scheme using bagasse in Minas Gerais (own source) 181

Table 6.4: Biomass projects implemented by UFPA in two communities (own source) 189

1 Introduction

One of the great challenges facing humanity during the twenty-first century must surely be that of supplying populations with safe and clean energy. Energy is a vital ingredient for economic growth, but energy production and use contribute to global warming. The majority of the gases which contribute to the greenhouse effect is directly associated to the production, conversion and consumption of fossil fuel energy. Hence, the question is how to meet rising energy demand and at the same time reduce environmental impacts.

The foreseen increase in energy demand is demonstrated in consumption trends. According to the International Energy Agency, energy consumption will have increased by 60% by 2030 (IEA, 2008). This increasing demand will have to be met by a mix of energy resources in order to supply energy needs, whilst considering environmental and socio-economic constraints.

In Brazil, hydropower accounts for approximately 77% of the energy supply (MME, 2009). However, the drastic changes in precipitation regimes during the last years are responsible for electricity shortages throughout the country. Moreover, energy consumption in Brazil increased considerably (MME, 2009). Hence, to guarantee the supply of electricity, a diversification of energy generation is necessary, prioritising technologies that use regional energy resources at a viable cost. In February 2010, a decree was issued to promote the generation of electricity from biomass through auction schemes (MME, 2010). The aim of this legislation is to include the electricity produced at decentralized biomass power plants in the distribution grids.

In Brazil, electricity consumption is concentrated in urban areas. During the last decades, the government has been implementing measures in order to extend rural electrification and diversify the energy matrix, which is mainly based on hydropower (ANEEL, 2009). Rural population in Brazil is located in remote areas, making the extension of the grid costly. In these areas, diesel generators often supply household electricity. This type of conversion is costly, sporadic and inefficient, hindering the utilization of energy for productive purposes or for vital tasks (Zerrifi, 2008).

Brazil is the world's largest exporter of sugar, ethanol, wood, coffee, orange juice and tobacco. The country's availability of land, water, and labour has increased production and exports during the last decades. Continuing expansion of trade and diversification of markets remain at the core of Brazil's agricultural growth strategy (Valdez et al., 2006). Brazil is a tropical country that produces a considerable volume of residues associated with large-scale agricultural activities. These biomass residues have the potential to influence positively the existing energy schemes if used as solid biofuels for energy generation. Energy use efficiency increase and co-generation of heat and electricity from local biomass residues could have a broad effect on the Brazilian environment and economy.

There are several conversion technologies for biomass, based on the type and availability of residues, as well as market demand. Various methods are used to turn raw materials into liquid, solid or gaseous energy sources. Biomass can be burned in a power plant to produce heat, fermented in an anaerobic digester to make biogas and then electricity and heat, or converted into a synthetic gas and fuel by thermo-chemical gasification. Which variant should be applied depends not only on the cost and expenses involved but also on the prevailing political conditions, priorities, needs and aspirations (FNR, 2009).

Fostering renewable energy handling and marketing is of vital importance. In many European countries, biomass residues have a consolidated market chain, which involves several stakeholders and processes. Nowadays, wood residues are more commercialized than crop residues due to the properties of the biofuel and the available industry norms.

Countries with broader experience in conversion technologies for biomass could encourage and support developing countries. Hence, the author found it relevant to investigate how agricultural and forestry residues could be used for decentralized energy generation in Brazil. Biomass samples of some agricultural crops were collected in Brazil and transported to Germany. Experiments at the BTU Cottbus were performed in order to investigate residues like rice and coffee husks, sugar cane bagasse and sawdust. Data based on the processing, compaction, combustion technologies and economic feasibility of Brazilian biomass residues are available in this thesis.

One of the challenges of this work was to reach reliable data in all stages of the production chain of biomass residues, in a branch where most residues have no commercial value and only scarce energetic use. Even possessing a flourishing ethanol industry and an extensive potential to obtain raw material from agricultural residues, Brazil has still little research on the establishment of supply chains of biomass. Moreover, a practical application of technology into rural areas is needed. Prior studies have concentrated on specific cases, mostly scoped in technology or end-use. This study proposes a set of variables and views in order to draw a realistic scenario for Brazil. To this end, the author visited several farmers and crop processing enterprises. In order to collect real data and understand how biomass residues could be included in the market, interviews were performed.

Energy supply and consumption should be addressed through integrated research. In this work, the author investigated issues related to resource management, environmental impacts, economic growth, energy security, climate change, employment, social welfare, investments and trade of Brazilian biomass residues. It was discovered that the implementation of biomass technologies and the improvement of socio-economic conditions in Brazil is mostly depending on energy policies, education and the promotion of innovation in further projects.

1.1 Developments in biomass research

Climate change issues have been fostering research and technological developments towards renewable energy sources. After the environmentalism wave of the 1970, most of European countries began to concentrate efforts in the development and implementation of so-called "green energy" technologies. Lately, scientists from various countries have been seriously engaging themselves in the promotion of energy generation from solid biomass.

The conversion of biomass into electricity has been studied in small-scale decentralized power plants (Thek and Obernberger, 2002). There have been few examples of successful projects that have been sustainable and/or replicated on a larger scale (Zerriffi, 2008).

The conversion of biomass residues into energy is a challenge for many scientists. Several German universities supported doctor theses related to the characterization and sensitivity analysis of solid biomass (Gamba, 2008 and Puttkamer, 2005). In order to reduce the technical constraints of biomass firing, Khan et al. (2009), Obernberger (1998) and Werther et al. (2000) have been investigating the influence of the ash melting point on the technical maintenance of furnaces.

The Technical University of Hamburg-Harburg (TUHH), with the support of the German Biomass Research Centre (DBFZ) and the European Union (EU), have been carrying out the project "BioNorm II - Pre-normative research on solid biofuels for improved European standards", which aims at the development of a market for solid biofuels within European standards. This project involves several European partners from Scandinavia, Austria and other countries (BioNorm, 2005), which traditionally use forest resources for energetic purposes.

The University of Lund in Sweden has been developing several projects regarding district heating systems. The project "Policy development for improving RES-H/C1 penetration in European Member States (RES-H Policy)" is supported by the European Commission through the IEE programme (RES-H Policy, 2010). On the European level the projects assesses options for coordinating national policy approaches in several EU countries. This results in common design criteria for a general EU framework of RES-H policies and an overview of the costs and benefits of various strategies.

As reported by Castro and Dantas (2009), Brazil started using biomass for energetic purposes in the late 1990's with the introduction of co-generation sugar cane ethanol plants. Universities in South Brazil started research and pilot projects in the field of co-firing using wood, rice husks and coal (Pereira et.al 2009). Professor Bazzo and Professor Schneider supervised postgraduate projects about coal and biomass co-firing (Finger, 2001 and Restrepo, 2010). Moreover, Professor Labate supervised in 2010 a doctoral thesis about the evaluation of the potential use of eucalyptus bark for energy production (Bragatto, 2010).

During the last decades, the use of ethanol and sugar cane bagasse in Brazil for fuel production and electricity generation increased from 10 to 51% (EPE, 2009). Most of the available documents cover mostly research in sugar cane cultivation, ethanol production, economics, logistics, marketing and energetic balance and the combustion

---

[1] RES-H: Renewable Energy Sources - Heating

of sugar cane bagasse for co-generation (Cunha, 2009; Kanadani, 2010; Molero, 2009; Silva, 2010 and Sosa, 2007). The author found few documented research and single official reports regarding the compaction of bagasse and other residues.

Nevertheless, the Federal University of Campinas (UNICAMP), located in São Paulo already displays excellence in sugar cane research and is taking some first steps in experimenting with biomass briquetting (Felfi et.al., 2010). The gasification of sugar cane residues and co-firing bagasse with natural gas have been deeply investigated at UNICAMP (Souza, 2001; Rodrigues et.al 2006 and Walter et.al 2008). The Federal University of Itajubá (UNIFEI) investigates the technical aspects of biomass such as gasification, combustion and pyrolysis. At UNIFEI, Prof. Lora (Lora and Salomon, 2005 and Lora and Andrade, 2009) performed research regarding the characterization of particulate emissions of sugar cane bagasse furnaces (Teixeira, 2005) and also a technical-economic approach to sugar cane production (Seabra, 2008). Data concerning social welfare and labour market perspectives in the sugar cane fields are also currently available (Azanha, 2007; Bartocci, 2009; Cardoso, 2010 and Gomes, 2010).

As showed by German (Kaltschmitt et al., 2009) and Brazilian scientists (Lora and Andrade, 2009), research concerning the use of biomass as an energy source has concentrated on technological aspects. The literature is mostly focused on the specific analysis of a particular technology. Project reports describe often only a particular activity implemented in a specific village or region, often neglecting the influence of energy policies and economy in the execution of such projects. Nevertheless, few transdisciplinary and holistic analysis of decentralized energy systems in rural areas of developing countries has been performed by authors like Goldemberg (2004), Cabraal et al. (2005), Chaurey et.al (2004), Hansen and Bower (2003), Paz et al. (2007) and Zerriffi (2006 and 2008) and institutions like the World Bank (1996) and ESMAP (2005).

1.2 Relevance

Quality of life and economic sustainability depend strictly on access to regular electric energy. Such access is a key element for the economic development and social welfare of rural areas (Pereira et al., 2011). Education and health quality depend on the provision of electricity.

Because most of the environmental and social benefits of bioenergy are externalities not considered in the priced market, leaving its development solely to the private sector will lead to economically efficient outcomes that may not, however, match their environmental and social potential (US AID, 2008). Therefore, concerns have been raised about the social implications of biomass use for energy generation and to what extent the biomass business could promote sustainable development.

The challenge of this thesis was to investigate all stages of the biomass production and utilization chain, from cradle to grave. A vast and rich literature helped the author identify each issue separately. The analysis of European literature was crucial for describing the technological state-of-the-art of biomass research and development. Market penetration was also assessed in order to create a scenario for Brazil. However, the biggest challenge was to collect data about Brazil.

The importance of this research relies on the assumption that the use of compacted biomass from Brazilian agricultural and forestry residues might be a cost effective alternative to partly replace fossil fuels while taking advantage of the resources available in the region to produce energy. An extensive utilization of heat and power generation using biomass might facilitate the provision of cleaner, more efficient energy services to contribute to local development, uphold environmental protection and improve livelihoods in rural areas.

By using local energy resources, employment can be created and as well perspectives for the local population. Furthermore, the increase of public acceptance for renewable technology use and the rural electrification are one of the most effectual tools to fight poverty. The existing infrastructure of conventional energy supply can be amended by decentralizing energy generation, thereby ensuring a sustainable energy supply.

The findings presented in this thesis reveal Brazil as land of possibilities in the energy sector. However, it should be considered that each biomass resource and its utilization chain are strictly dependent on the socio-economic and cultural characteristics of the regions where is produced. This research has several implications. The methodology used by the author, linking technology with socio-economic aspects and experimental activities with personally communication in a transdisciplinary way, could serve as inspiration for other researches.

1.3 Objectives

The present research carried out detailed field work in order to evaluate the energetic potential of Brazilian biomass in various regions. This potential could foster research towards regional use of biomass for decentralized energy generation in rural areas. Here, the author identifies the basic needs, typical constraints and probable challenges of rural electrification and economic development in Brazil. Several case studies and social groups were investigated in the Federal States of Minas Gerais, São Paulo, and Pará, regarding industrial processes, lightning, heating and cooking.

The aim was to understand individuals and communities, living in isolated areas of the Amazon region and non-isolated areas in Southeast Brazil. They presented different social, cultural and geographical living conditions, which were mostly related to access to electricity. The author identified two main concerns regarding the utilization of biomass residues in Brazil:

1. Technological challenges:
    - State-of-the-art of technologies
    - Compaction of biomass residues and combustion behaviour
2. Socio-economic and institutional issues:
    - Economic feasibility
    - Decentralized energy systems

The aim was to approach biomass technologies and identify the institutional issues in Brazil impacting rural electrification. The author investigated regulations, policies and access to financing. She evaluated the effects of policies on the development of initiatives for decentralized energy generation using biomass resources, taking into account the socio-economic and geographical peculiarities of the different Brazilian regions.

The properties of Brazilian biomass were investigated through a selection of the main harvested crops like sugar cane, eucalyptus wood, coffee and rice. Some properties of these residues were analysed at the Chair of Power Plant Technologies and the Chair of Mineral Processing, Processing, and Refinement of Biogenous Resources. The parameters first investigated were moisture, ash content, calorific value, and ash melting point.

The specific objectives of this work can be summarized as follows:

- To assess the potential of agricultural and forestry residues in Brazil through evaluation of existing studies and exchange of knowledge and information with Brazilian partners
- To assess how residues are gathered and used: logistics in Brazil, storage and transportation of biomass
- To characterize Brazilian residues: physical-chemical properties and compaction
- To determine the socio-economic feasibility of pellet and briquette production for decentralized energy generation in Brazil
- To investigate how stakeholders could upgrade the biomass residues, generate local heat and power and sell the compacted biomass and surplus electricity
- To outline strategies to increase energy use efficiency, resource upgrading, and commercialization.

1.4 Methodology of work

Case study sampling was done in combination with review of the literature on Brazilian biomass resources and the generation of residues in agriculture and forestry activities. Also, a detailed overview of the state of the art of biomass technologies was necessary in order to widen the scope of possibilities for energy conversion using local resources. In this sense, the author could understand how technology implementation might or might not be feasible for each investigated region in Brazil. By reading the Brazilian energy policies, the author could interpret the meaning of the several institutional modifications and its effects on the economic development of the country, especially in rural areas.

The selection of case studies was initially based on the availability of biomass for sampling and further laboratory experiments. The author collected and transported to Germany biomass samples, which originated from the main harvested crops: sugar cane, rice, coffee and eucalyptus sawdust. This was only possible with the logistic and legal support of the Federal University of Viçosa, located in the Federal State of Minas Gerais.

The investigation of biomass properties was done at BTU Cottbus laboratories. The author demonstrated the technical feasibility of the residues for compaction, combustion and conversion into energy in two published articles (Missagia et al. 2010 and 2011). However, the author still needed to learn how biomass is used for energetic purposes in Brazil and where there exists a market for biomass residue. Travelling from South to North, she could gather information essential to this research.

The idea of expanding the geographical scope to North Brazil was due to findings regarding the commercialization of residues and the current electrification policies of the country. The author sought to access areas where electricity is almost universally provided (Southeast Brazil) and where there is a lack of grid connection due to the remoteness (North Brazil). Through field investigation and the interviewing of experts in the biomass business and farmers, the author could extend her assumptions regarding biomass use in Brazil.

1.5 Structure of the thesis

The author performed exploratory measures and feasibility studies that cover the technical, socio-economic and environmental aspects of biomass utilization in Brazil. This was done through extensive literature research, including academic publications such as scientific journals, reports and PhD theses. The choice of types of residues to be investigated was based on the biomass availability in the region where the partner university is located. Field excursions and interviews were critical to understanding land cultivation and the perspectives of farmers.

Biomass cultivation in Brazil and the utilization of sub-products for energy generation will be described in **Chapter 2**. The implication of the current energy policies for the extension of biomass projects is also a crucial issue in this chapter.

The state-of-the art of the technological advances in biomass conversion are depicted in **Chapter 3**. The physical-mechanical and chemical properties of biomass inform the feasibility of compaction and combustion for different residues.

The pelleting of Brazilian biomass was conducted in pilot scale for sugar cane bagasse, eucalyptus sawdust, rice, and coffee husks. The experiments and the results are described in **Chapter 4**.

Chapter 5 explores the economic feasibility of rural decentralized energy systems based on compacted biomass residues.

In **Chapter 6**, the author aimed to explore the regional context in order to identify a set of prospects and challenges for rural communities in Southeast and North Brazil that could shape possible development trajectories for the utilization of biomass residues aiming energy generation.

1.6 References

ANEEL – Agência Nacional de Energia Elétrica. Brazilian Energy Matrix 2009 Available at: http://www.aneel.gov.br, Retrieved on: 20 April 2010.

Azanha M. (2007) O mercado de trabalho da agroindústria canavieira: desafios e oportunidades (in Portuguese). Departamento de Economia, Administração e Sociologia – ESALQ/USP – Piracicaba, SP. Econ. aplic., São paulo, v. 11, n. 4, p. 605-619.

Bartocci L. (2009) Perfil da mão de obra no setor sucroalcooleiro: Tendências e perspectivas. Tese de Doutorado: Programa de Pós-Graduação, Departamento de Administração. Faculdade de Economia, Administração e Contabilidade, Ribeirão Preto – USP

BioNorm (2005) Pre-Normative Work on Sampling and Testing of Solid Biofuels for the Development of Quality Assurance Systems. Final technical report. Available at: http://www.bionorm2.eu/downloads/KeyDataandExecutiveSummary.pdf

Bragatto, J. (2010) Avaliação do potencial da casca de Eucalyptus spp. para a produção de bioetanol. Tese de Doutorado: ESALQ - Luiz de Queiroz College of Agriculture, Plant Physiology and Biochemistry, Brazil

Cabraal RA., Barnes DF. and Agarwal SG. (2005) Productive uses of energy for rural development. In: Annual Review of Environment and Resources 30: 117-144

Cardoso T. (2010) Cenários tecnológicos e demanda da capacitação da mão-de-obra do setor agrícola sucroalcooleiro paulista (in Portuguese). Dissertação de Mestrado: Universidade Estadual de Campinas (UNICAMP), Faculdade de Engenharia Agrícola.

Castro NJ. and Dantas GA (2009) Fusoes e aquisicoes no stor sucroenergético e a importância da escala de geracao. In: TDSE n.14, Rio de Janeiro. Available at: http://www.nuca.ie.ufrj.br/gesel/tdse/TDSE14.pdf (Retrieved on February 1, 2011)

Chaurey A., Ranganathan M. and Mohanty P. (2004) Electricity access for geographically disadvantaged rural communities: technology and policy insights. In: Energy Policy 32: 1693-705. Elsevier

Cunha J. (2009) A estrutura socioeconômica da produção de etanol no Brasil: o uso de fatores primários de produção e as suas relações intersetoriais. Tese de Doutorado: ESALQ - Luiz de Queiroz College of Agriculture, Applied Economics.

EPE - Empresa de Pesquisa Energética (2009) Balanço Energético Nacional 2009: Ano base 2008 / Empresa de Pesquisa Energética. – Rio de Janeiro: EPE, 2009.

ESMAP – Energy Sector Management Assistance Program (2005) Brazil background study for a national rural electrification strategy: aiming for universal access. Washington DC. Available at: http://vle.worldbank.org/bnpp/files/TF027963ESMA06605BrazilFinal.pdf (Last access 6, Mai 2011)

Felfi FF., Mesa JM., Rocha DJ., Filippeto D., Luengo CA. and Pippo WA. (2010) Biomass briquetting and its perspectives in Brazil. In: Biomass and Bioenergy (in Press)

Finger C., Schneider P., Bazzo E. and Klein J. (2001) Efeito da intensidade da desrama sobre o crescimento e a produção de Eucalyptus saligna. Cerne 7(2): 53-64. Available at: http://www.dcf.ufla.br/cerne/Revistav7n2-2001/06%20artigo%20016.pdf (Last access 6, January 2011)

FNR – Fachagentur für Nachwachsende Rohstoffe (2009) Bioenergy. 2nd Edition http//:www.fnr-server.de/ftp/pdf/literatur/pdf_330-bioenergy_2009.pdf

Gamba L. (2008) Erste Modellentwicklung zur nachhaltigen Nutzung von Biomasse. Doctor thesis. Fakultät Prozesswissenschaften, TU Berlin.

Gomes J. (2010) O canavial como realidade e metáfora: leitura estratégica de trabalho penoso e da dignidade no trabalho dos canavieiros de Cosmópolis (in Portuguese). Institute of Psychology, Social Psychology, USP.

Hansen CJ. and Bower J. (2003) An economic evaluation of small-scale distributed electricity generation technologies. In: Oxford Institute for Energy Studies, Oxford.

IEA – International Energy Agency (2008) World Energy Outlook 2008. Available at: http://www.iea.org/textbase/nppdf/free/2008/weo2008.pdf

Kanadani (2010) Fundamentos econômicos da formação do preço internacional de açúcar e dos preços domésticos de açúcar e etanol (in Portuguese). Tese de Doutorado: Luiz de Queiroz College of Agriculture, Applied Economics.

Kaltschmitt M., Hartmann H. and Hofbauer H. (2009) Energie aus Biomasse: Grundlagen, Techniken und Verfahren; Springer, Berlin, 2009, 2. Aufl.

Lora ES. and Salomon KR. (2005) Estimate of ecological efficiency for thermal power plants in Brazil. In: Energy Conversion and Management 46: 1293-1303 Elsevier

Lora ES. and Andrade RV. (2009) Biomass as energy source in Brazil. In: Renewable and Sustainable Energy Reviews 13: 777-788 Elsevier

Missagia B., Guerrero C., Narra S., Krautz H J., Ay P. (2010) Physical characterization of Brazilian agricultural and forestry residues aiming the production of energy pellets. In: 18th European Biomass Conference and Exibition, Lyon.

Missagia B., Corrêa MF., Ahmed I., Krautz HJ. and Ay P. (2011) Comparative analysis of Brazilian residual biomass for pellet production. In: Implementing Enviromental and Resource Management (Eds. Schmidt M., Onyango V. and Palekhov D.) Springer Verlag Berlin Heidelberg. ISBN 978-3-540-77567-6

MME – Ministério de Minas e Energia (2009) Resenha Energética Brasileira 2009. Available at: http://www.mme.gov.br/mme/galerias/arquivos/publicacoes/BEN/3_-_Resenha_Energetica/Resenha_Energetica_2009_-_PRELIMINAR.pdf (Retrieved on January 11, 2011)

MME – Ministério de Minas e Energia (2010) Portaria No 56, de 4 de Fevereiro de 2010 Available at: http://www.mme.gov.br/mme/galerias/arquivos/noticias/2010/Port_56_Leilxo_Biomassa_nos_Sistemas_Isolados.pdf (Retrieved on November 1, 2010)

Molero L. (2009) Modelo de formação de preços de commodities agrícolas aplicado ao mercado de açúcar e álcool (in Portuguese). Tese de Doutorado: Programa de Pós-Graduação, Depto. de Administração. Faculdade de Economia, Administração e Contabilidade USP.

Paz LR., Fidelis NS. and Pinguelli RL. (2007) The paradigm of sustainability in the Brazilian energy sector. In: Renewable and Sustainable Energy Reviews 11: 1558-1570

Pereira LF., Bazzo E., Oliveira Jr. and Martins A. (2009) Biomass Co-firing as an Alternative Technology for a Clean Coal Electric Generation in Brazil, 20th International Congress of Mechanical Engineering, Gramado, Brazil, ABCM

Pereira MG., Freitas MA. and Silva NF. (2011) The challenge of energy poverty: Brazilian case study. In: Energy Policy 39: 167-175 Elsevier

Puttkamer T. (2005) Charakterisierung biogener Festbrennstoffe. Doctor thesis. Fakultät Maschinenbau, Universität Stuttgart.

RES-H Policy (2010) Policy development for improving RES-H/C2 penetration in European Member States. Available at: http://www.res-h-policy.eu/ (Retrieved on November 1, 2010)

---

[2] RES-H: Renewable Energy Sources - Heating

Restrepo (2010) Metodologia de análise exergoambiental para plantas termoelétrica operando no processo co-firing carvão – biomassa (in Portuguese). Programa de Pós-Graduação em Engenharia Mecânica Centro Tecnológico, Universidade Federal de Santa Catarina.

Rodrigues M., Walter A. and Faaij A. (2007) Performance evaluation of atmospheric biomass integrated gasifier combined cycle systems under different strategies for the use of low calorific gases. In: Energy Conversion and Management 48: 1289-1301 Elsevier

Sales CVB., Andrade RV. and Lora EES. (2006) Geração de eletricidade a partir da gaseificação de biomassa. In Proceedings of the 6. Encontro de Energia no Meio Rural, 2006, Campinas - SP, Brazil (Retrieved on January 10, 2011) Available at: http://www.proceedings.scielo.br/scielo.php?script=sci_arttext&pid=MSC0000000022006000100067&lng=en&nrm=iso

Seabra JEA. (2008) Avaliação técnico-econômica de opcoes para o aproveitamento integral da biomassa de cana no Brasil. Doctor thesis submitted to the Federal University of Campinas, Brazil.
Available at: http://cutter.unicamp.br/document/?code=vtls000446190&fd=y (Retrieved on November 1, 2010)

Silva A. (2010) Desenvolvimento de fatores de normalização de impactos ambientais regionais para avaliação do ciclo de vida de produtos no Estado de São Paulo Tese de Doutorado: School of Engineering of São Carlos, Environmental Engineering Sciences USP.

Sosa (2007) Caldeiras aquatubulares de bagaço : estudo do sistema de recuperação de energia
Universidade Estadual de Campinas (in Portuguese). Tese de Doutorado: Programa de Pós-Graduação em Engenharia Mecânica, UNICAMP.

Souza MR. (2001) Co-firing como alternativa para impulsionar a tecnologia de gaseificacao de biomass integrada a ciclos combinados – BIG-CC. Tese de Doutorado: Programa de Pós-Graduação em Engenharia Mecânica, UNICAMP.

Teixeira FN. (2005) Controle da emissão de particulados em caldeiras de bagaço. Doctor thesis submitted to the Federal University of Itajubá, Brazil. Available at: http://www.nest.unifei.edu.br/portugues/pags/teses/conc_doutorado_mec/files/controle_ da_emissao_de_particulados_em_caldeiras_para_bagaco.pdf (Retrieved on November 1, 2010)

Thek G. and Obernberger I. (2002) "Wood pellet production costs under Austrian and in comparison to Swedish framework conditions" In: Proceedings of the 1st World Conference on Pellets. Swedish Bioenergy Association (ed), Stockholm, Sweden:123-128.

Valdez C., Lopes IV., Lopes MR., Oliveira MS. and Bogado PR. (2006) "Factors Affecting Brazilian Growth or are there Limits to Future Growth of Agriculture in Brazil?" 20 Nov. 2008 Available at: http://www.ers.usda.gov/AmberWaves/November06/ (Retrieved on November 1, 2010)

Walter A., Dolzan P., Quilodrán O., Garcia J., da Silva C., Piacente F., Segerstedt A. (2008) A Sustainability Analysis of the Brazilian Bio-ethanol. Unicamp. Department for Environment, Food and Rural Affairs (DEFRA), British Embassy, Brasilia.

Wether J., Saenger M., Hartge E., Ogada T., Siagi Z. (2000) Combustion of agricultural residues. In: Progress in Energy and Combustion Science Pp. 1-27, Elsevier Pergamon

Zerriffi H. (2006) Making small work: serving rural electricity needs on a local scale. In: Liu Institute Newsletter 3. Draft manuscript

Zerriffi H. (2008) From acaí to access: distributed electrification in rural Brazil. In: International Journal of Energy Sector Management 2: 90-117 Emerald Group Publishing Limited

2 Biomass and energy generation in Brazil

Brazil is a vast country with a strong regional distinction, varying geographically, culturally and economically. A centralized energy system with a National Grid supplies 174 million of people with electricity (ANEEL, 2008). The remaining 10 million Brazilians, living mostly in the remote areas of the Amazon and in rural areas, are supplied by decentralized, isolated energy systems. The difficulties in connecting North Brazil with the National Grid arise mainly out of the peculiarities of the region, which has dense forests, large rivers, low population density and low income.

Brazil's electricity supply relies on hydropower. Because of the strong reliance on large hydropower dams for electricity generation, most households are supplied by a central grid transmission system (ESMAP, 2005). But due to the environmental impacts caused by landscape flooding, the government began restricting the construction of new hydropower plants. This allows the introduction of other renewable energy sources such as biomass and the diversification of the Brazilian energy sector.

Producing more energy from biomass residues has development potential in Brazil. Agriculture accounted for 6% of the country's gross domestic product (GDP) in 2010 (Department of State, 2011). Brazil's worldwide success in sugar cane production for ethanol has opened new possibilities for the use of agriculture residues. Steam generation from sugar cane bagasse and leaves could supply inhabitants with as much electricity as Brazil's largest hydropower plant Itaipú (MME, 2008).

With 5 % of the total energy generated (ANEEL, 2009), biomass electricity has been promoted through investments, new policies and regulations, and trough agricultural and technological developments. Sugar cane bagasse represents 73.5% (2.926 MW) of the total installed power from biomass power plants. Further biomass sources used in power plants such as black liquor, rice husks, wood residues, and charcoal contribute little to energy generation in Brazil.

2.1 Historical background

In Brazil, the organization and production model before 1500 was based on indigenous collective subsistence (Ribeiro, 1997). The period of the Portuguese colonization brought slavery and genocide. Within a few decades many indigenous tribes had vanished from the Brazilian coast.

From the 16th to 19th century, the colonial economy was based on slavery and on the export of timber, sugar cane, gold, tobacco, cotton, and coffee. Each of these activities had their economic up and downs. The stagnation of the 19th century's economy was due in large part to the collapse of gold mining. The end of the 19th century saw the abolition of slavery in Brazil. In the beginning of the 20th century, the distribution of wealth in the country was directly related to the agricultural production and export of tobacco, cotton, and coffee. The southern Federal States had more capital and a considerably larger consumer market compared to the northern regions.

Industrialization in Brazil occurred during the 1950s. From the 1950s until the beginning of the 1990s, the Brazilian dictatorship considered that progress and efficient energy supply was best reached through the construction of massive hydropower plants and centralized grid connections. Electricity production takes place far from the largest consumer markets, located in the southern regions (Walter and Dolzan, 2009). In contrast, current environmental legislation restricts the implementation of such hydropower projects in the north, due to their significant environmental impacts in the Amazon region.

During the 1960s and 1970s, the modernization of agriculture began in Brazil. The propagated use of chemical fertilizers, pesticides and machinery made possible the cultivation of great land extensions. In the beginning of the 1970s, the Oil Crisis pushed the Brazilian government to invest in the sugar cane ethanol sector. Farmers were persuaded to plant sugar cane in a monoculture system.

The drastic switch from small-scale to large-scale farming models during the 1970s resulted in environmental, economical and social problems in Brazil. In the mid 1960s a project for the Brazilian agriculture was launched aiming to modernize the structure of agricultural production in order to be competitive with developed countries. Fast and complex changes in agricultural production had negative social and environmental impacts. The use of industrial inputs such as fertilizers, pesticides became widespread. By the 1970s mostly large-scale farmers were benefited (Martine, 1987).

In the early 1990s, the government removed much of the State's intervention in the agricultural business. With liberalized trade and strengthened market signals, imports and use of agricultural inputs and technology increased markedly throughout the 1990s. The improved investment climate and reduced border controls also ushered in foreign direct investment, which increased competitiveness and efficiency in the agricultural sector (Schnepf et al. 2001). Since the mid-1990s, the most significant economic factor has been the introduction of a successful economic stabilization plan.

Brazil has a long history of economic instability involving hyperinflation, overvalued exchange rates, and frequent currency realignments. In the mid-1980s, economic reforms eliminated domestic and export taxes and restrictions on agriculture. Export taxes and quotas were used extensively to dampen internal prices and encourage domestic processing, while high tariffs and import controls on agricultural inputs promoted "import substitution" programs benefiting domestic industries. The reforms contributed to a more stable macroeconomic environment for investment and decision-making (Schnepf et al. 2001).

Today, Brazil is an important food producer and supplier of international markets, being the main exporter of sugar, soybeans, ethanol, coffee, orange juice, timber and tobacco. The country's availability of land, water, and labor makes Brazil a naturally low-cost producer of crops. One third of the potential arable land is used and the expansion of agricultural areas depends on environmental and social impact assessment, access to credit for production, investment in infrastructure, and transport.

2.2 The potential of biomass in Brazil

The current political scenario in Brazil supports the introduction of combustion technologies for biomass, mainly due to the great expansion of sugar cane ethanol production using bagasse, together with the generation of bagasse electricity. Other kinds of residues like wood sawdust and rice husks also have a great potential either for covering the local energy demand or for commercialization in the form of pellets or briquettes.

One of the main advantages of such conversion process is allowing the technology to be optimized according to technical, environmental, and socio-economic criteria referring to the special fuel properties and regional differences (Kaltschmitt et al. 2004). Small-scale systems could supply a reliable, clean and economically feasible energy source to rural communities when supplied with biomass from local resources. Moreover, according to the Alliance for Rural Electrification (ARE, 2010), bioenergy systems are regarded as low investment, considering mechanization and operation, which might result in high local labor, demand and low investment rate per local job created compared to other energy technologies.

The literature often describes the production and use of agricultural residues according to various Brazilian regions, demand, and local business (Filho et al., 2004 ; Goldemberg, 2007, Lora et al., 2009 and Nass et al., 2007). In general, it is assumed that about 25% of any dry agricultural feedstock is a residue (Eriksson and Prior, 1990). Table 2.1 shows the residues of some Brazilian crops.

Table 2.1: Main sources of agriculture residues in Brazil (Couto el. al., 2004)

Crops	Residue percentage in crops (%)
Cotton	50
Rice	20 to 25
Coconut	50
Coffee	20
Sugar cane	80
Sunflower	10 to 15
Corn	25 to 30

The following chapters will depict some of the main important Brazilian products: rice, coffee, sugar cane and timber. Samples of each mentioned biomass were collected in the rural area of the city of Vicosa, in the Federal State of Minas Gerais. The author chose these crops for their feasibility of sampling, the high availability throughout the country, and the assumption that the residues could be a potential raw material for compaction and combustion. In chapter 4, the physical-chemical characteristics of these residues will be presented.

2.2.1 The coffee tradition in Brazil

In the end of the 19th century, coffee farms started to be established in Brazil, being located mainly in the Southeast region. In the first years of 1910s, the Federal State of São Paulo was producing more than half of world's coffee. Between 1890 and 1930, during the so-called 'Coffee Era', the cultivated coffee made up the majority of the country's exports. The 'Coffee Era' played a crucial role in revitalizing Brazil as an agro-exporter in the world market and as a potential importer of industrial goods. After the First World War, the surplus production of coffee, the loss of British interest in buying Brazilian coffee, and the depletion of arable soil were important factors which caused the end of the Coffee Era in São Paulo.

In 2010, about 2.75 million tons of coffee was produced in Brazil (IBGE, 2011). Today, the Federal State of Minas Gerais, located in the Southeast, is the lead exporter of coffee, with almost 52% of national production.

The processing of one ton of coffee generates approximately 20% of residues. Therefore the amount of residues generated would be around 548 thousand tons per year. The majority of coffee farmers employ part of this residue directly in the soil. This practice aims soil protection against erosion and fertilization. Even when coffee husks have satisfactory chemical composition in comparison to other organic fertilizers, mainly regarding N and K contents, the husk is bulky, making storage, handling and soil incorporation a problem (Mazzafera, 2002). Additionally, coffee husk transportation for organic fertilization represents high costs for the farmers, especially in hilly areas.

2.2.2 Sugar cane and ethanol

Sugar cane is a crop that has been cultivated in Brazil since the colonization. Today, sugar cane is a highly profitable and competitive crop within the overall agribusiness sector of Brazil (Shikida, 2010). Brazil is the most important sugar cane producer of the world, followed by India, Thailand and Australia, and is responsible for 45% of world-wide ethanol production (UNICA, 2008).

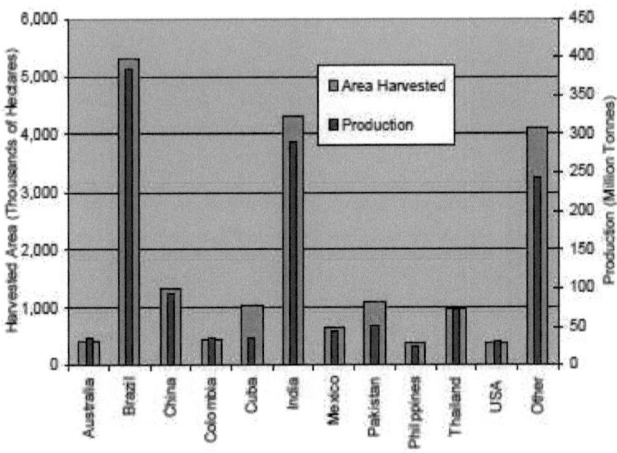

Figure 2.1: Sugar cane production by country (FAO in WADE, 2004)

Brazil achieved worldwide importance in ethanol production as a result of a combination of factors such as favorable weather, cheap labor, and extensive governmental support (Bourne, 2007). The Brazilian government started supporting ethanol production as a strategy to the 1973 oil crisis. The decision to produce ethanol from sugar cane was based on the low cost of sugar at the time and the country's tradition and experience with this feedstock. The program Pró-Álcool was created in 1975 and sponsored by the government to phase out transportation vehicle fuels derived from fossil fuels in favor of sugar cane ethanol (Briquet, 2007). The military dictatorship at the time heavily subsidized and financed new ethanol plants and the state-owned oil company, PETROBRÁS, to install ethanol tanks and pumps around the country (Bourne, 2007).

In 1988 sugar prices harshly augmented in the international market. There was a shift in ethanol production towards sugar production, resulting in an ethanol supply shortage (Briquet, 2007). Consequently, ethanol prices rose and the elimination of some ethanol subsidies led to a drop in ethanol supply (Ribando et al. 2007). This phase persisted until 1999, when a rise in oil prices resulted in a new demand for ethanol (Winfield, 2008). Besides the encouragement in production, the Brazilian government also offered a tax on sugar exports in 2002 whenever ethanol appeared in short supply to meet the demand in order to prevent the diversion of sugar cane to the international markets.

In 2003, flex fuel vehicles were launched in Brazil, which run on gasoline, ethanol, or a mixture of both. A federal law established mandatory blends of ethanol and gasoline. The government set the blend percentages based on the sugar cane harvest and the ethanol production, resulting in variations even within the same year.

Some sugar mill/distillery complexes are capable of producing sugar and ethanol using the heat and electricity generated from bagasse combustion. The generation of electricity from sugar cane bagasse is as an option to seasonally complement the hydropower sector, in relation to the rainfall regimes. Since the 1990s the ethanol distilleries are operated with cogenerated electricity. The surplus can be sold to the grids. This is a competitive advantage of Brazilian ethanol since it can be produced without using electric power generated by other sources.

Bagasse electricity contributes to the reduction of GHG emissions. Moreover, the generation costs from large-scale bagasse plants already compete with the hydropower plants. Ethanol plants are located next to the main transmission grid in the State of São Paulo. This is also a decisive factor in diminishing the costs and the electricity losses with transmission. However, hydropower still guarantees lower prices to consumers (GESEL, 2011).

Between 25 % and 30% of the sugar cane production is bagasse (Rosillo-Calle et al. 2007). Ethanol represents only one third of the energy available in cane (Figure 2.2); the other two thirds represented by fibbers in the cane stalks (bagasse) and in cane leaves (straw) is almost fully used in the process in the following way:

- 93% of the bagasse is used as fuel in cane processing,

- 85% of the straw is burned directly on the fields to reduce the costs of harvesting. The other 15% is harvested unburned but the straw is left on the ground to decay. In both cases the fiber carbon returns to the atmosphere in the form of CO_2.

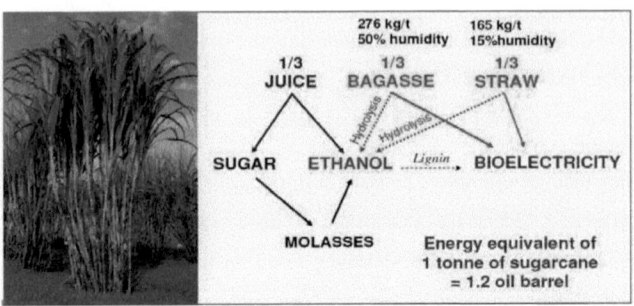

Figure 2.2: Sugar cane: source of energy (Source: UNICA in Jank, 2008)

In most ethanol and sugar plants of Brazil the bagasse has been burned for generating thermal and electrical energy since 1996. The heat is used in the distillation process and transformed into steam, which rotates the turbines generating electricity. The self-produced electricity is used in the plants and the surplus sold to the grids. The energy output from bagasse is largely dependent on moisture, technology and the electricity market rules (WADE, 2004).

According to WADE (2004), a typical sugar cane complex with a capacity of 3,000 tones of crushed cane per day can produce 345 tonnes of sugar, 6,000 litres of ethanol, 3 tonnes of yeast, 15 tonnes of potash fertilizer, 25 tonnes of pulp, 15 tonnes of wax, 150 tonnes of press-mud fertilizer and 240 MWh of exportable electricity from bagasse.

As reported by the author, an ethanol plant located in Ribeirao Preto, Brazil consumes more than 30,000 tons of bagasse per year in a 2 block - 15 MW power plant. The average cost to build a new plant is about BRL 2,800/kW, not considering the installation and connection costs to the distribution system.

The electricity generated is used for the ethanol production and the surplus is sold to the grids. In this plant, each ton of sugar cane bagasse produces 60 kWh of energy. More than one third of this bagasse is bought from other ethanol factories for 10 BRL/ton. Despite of the transportation costs of bagasse (20 BRL/ton), the electricity generated compensates the efforts. The total electric power is 46 MW/h. The ethanol production consumes 20 MW/h and the rest is sold to the national electric system. As reported, the price for 1 MW/h in Brazil is much higher for electricity generated in centralized hydro power plants (BRL 345.00 in São Paulo) than in decentralized ethanol plants (BRL 153.00).

However, the full exploitation of the sugar cane energy potential requires modernization of the plants. It is important to improve industrial processes to generate more bagasse. The harvest of unburned cane and therefore the recovery of cane straw could increase biomass volume (Hassuani et al. 2005). The surplus of 40 - 80 KWh/ton of processed sugar cane is related only to bagasse. The use of sugar cane straw for electricity generation could permit an estimated production of 200 KWh/ton of processed sugar cane (Castro et al. 2010).

The great advantage of the bioelectricity projects is the capability to cover the energetic demand when the water reservoirs are almost empty. The dry period coincides with the sugar cane harvest, which occurs from April until November (Figure 2.3). The hydropower reservoirs and the sugar cane plants are both located in Southeast and Center West Brazil. Hence, bioelectricity from bagasse could complement hydropower.

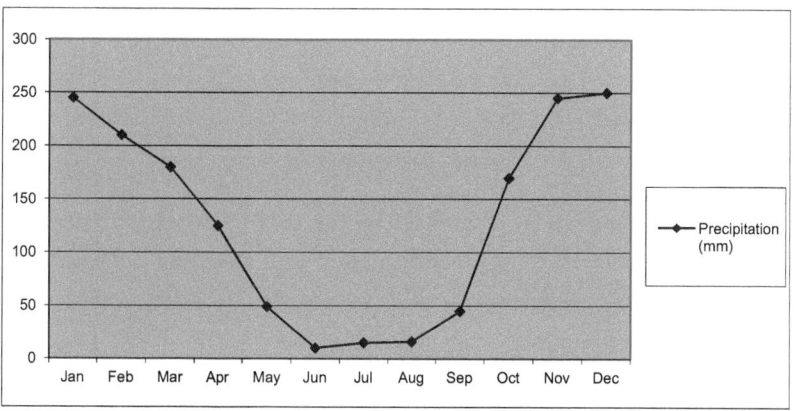

Figure 2.3: Average precipitation levels in Brazil (CPTEC, 2011)

During the last ten years, ethanol production has been growing with an increase of exports. According to the IBGE, between 1998 and 2008 the planted area of sugar cane increased almost 150% in the Center West region (Annex 1). It has been forecasted that the 2011 sugar cane harvest will reach almost 600 million tons, compared to 541.5 million tons last season. Of the total cane volume expected for the 2011 harvest, UNICA projects that 43% will be destined for the production of sugar, and almost 57% will be used for the production of ethanol (UNICA, 2011).

Brazil is today the world's largest exporter of ethanol, with 430 plants concentrated mainly in the Center-South of Brazil. About 60% of production comes from São Paulo and 10% from the Northeast. The Federal State of São Paulo produced ca. 325 million tons of sugar cane in the 2010 harvest and 16.84 billion liters of ethanol (GOB, 2011). The planted area of sugar cane in São Paulo has increased in the last years as shown in Table 2.2.

Table 2.2: Sugar cane planted area: Brazil and São Paulo (IBGE, 2011)

	Sugar cane planted area (1,000 ha)					
	2004	2005	2006	2007	2008	2009
Brazil	5,634	5,815	6,390	7,086	8,211	9,674
São Paulo	3,414.2	3,673.3	4,258.4	4,835.4	5,411.3	5,538.92

Sugar cane ethanol plants represent a path towards diversification of the Brazilian energy sector using biomass. The realization of bagasse electricity projects is simpler to manage because such plants have lower environmental risks. The ethanol plants have special procedures to treat their wastes and minimize the environmental impacts.

Today, the ethanol business is a success due to the opening of the electricity market to biomass power producers, the private ownership of sugar mills and the stimulation of rural activities based on biomass energy to increase rural employment. Through the years, the technological advancements and the generation of electricity from sugar cane bagasse have contributed to maintain Brazil's leadership position in biofuel production. Moreover, Brazilian industry is able to supply sugar cane factories with equipments, completely independent from imports.

The sugar cane agroindustry for ethanol production in Brazil is a significant job generator including a range of competences and training, though most of them are low skill jobs (BNDES / CGEE et al. 2008). On the other hand, there has been a sharp decrease in employment in the sugar cane sector as the result of the increase in mechanical harvesting in sugar cane production. It has been documented that one harvesting machine can replace up to 100 workers (Smeets et al. 2008). Alternatively, the positive employment effect from the use of sugar cane residues for the generation of energy could compensate the negative employment effect, by relocating the work force released by mechanical harvesting.

2.2.3 The rice industry in South Brazil

Rice cultivation in South Brazil began with the colonization and its rise as a national food. Today, the Federal State of Rio Grande do Sul is the core of irrigated crop acreage and produces an important share of Brazil's rice in an area of approximately 1 million hectares. The largest production and the largest rice yields are located in South Brazil. The production of rice in Brazil was about 7.7 million ton in 2010 (GOB, 2011).

Before the commercialization of rice takes place it is necessary to dry and peel the rice. There are 300 rice-processing plants in Rio Grande do Sul. Both steps do not occur necessarily directly near the cultivation fields. The transport of the raw rice (with husks) is economically feasible, since drying and peeling preferably mostly occurs near the consumers and the exporting handlers. The rice processing plants are classified in three groups: the white rice plants, the parboil rice plants and the joint white and parboil rice plants. The greatest part of the Brazilian production is of white rice in Rio Grande do Sul.

The rice is transported to the processing plants with 25 to 30% moisture, varying with type of cultivation and period of the year. The bulk density of rice husks is very low (see annexes). Hence, the transportation of this material is problematic. Many companies are compacting rice husks into briquettes to improve fuel characteristics (Embrapa 2010, Marcon et al. 2004 and FAO, 1990). The first briquetting machines in Brazil were sold to rice-mills in the 1980s. Afterwards, these machines were expanded to wood factories and sawmills with occasional sales to other agro-residue producers such as coffee mills (FAO, 1990).

The rice harvest occurs between January and April. Rice processing and storage demands specific moisture contents of 12 to 15%. Rice husks are fired in direct combustion furnaces, which produce heat and the flue gases are used to dry the rice.

Approximately 22% of rice is made up of husks and each year approximately 760 thousand tons are produced in Rio Grande do Sul. The rice husks take more than 2 years to decompose and expel methane gas, harmful to the environment. Therefore, most of the farmers in South Brazil burn up about 15% of the leftovers to dry the harvested rice and 35% are sold for bedding chickens (Eriksson and Prior 1990).

The processing facilities using rice husks are mostly located in South Brazil. Large-scale facilities like the enterprise CAMIL use the rice husks to generate heat for drying the rice and to produce electricity. The combustion of rice husks produces a high amount of ash but little sulphur (see chapter 4).

The CAMIL Biomass Electricity Generation Project in South Brazil has been operational since 2001. National and regional legislation prohibits the unlicensed displacement or uncontrolled burning or land filling of rice husks. Before the power plant began operations, 81% of rice husks were disposed in legal landfills. After October 2005, surplus electricity has been sold to the grid, using 93% of all generated rice husks for project activity. The surplus of 7% rice husks is disposed in legal landfills. The amount of CO_2 avoided because of this project is estimated to be 57,341 tonnes CO_2 per year and its certified under the Clean Development Mechanism (Clear, 2010).

As considerable amounts of rice husks are not currently utilized in small-scale farms in Brazil, the implementation of decentralized energy systems in rural areas could follow the success of CAMIL. In order to achieve that, small-farmers could developed supply chains of rice husks to provide the necessary feedstock for heat and power generation. It was documented by the author that in several small-scale farms which cultivate rice, the chicken farmers are collecting the husks.

2.2.4 *Forestry*

The exploitation of wood has been an important economic activity in Brazil since the times of colonization. Forestry is concentrated in the Amazon region (North States) and South, mainly in the Federal States of São Paulo, Minas Gerais, Paraná and Rio Grande do Sul. In the Amazon region the production is mainly based on timber falling, while in the South forestry is based on short-rotation trees.

2.2.4.1 Native forests

The regions that have traditionally exploited the wood resources, such as the Amazon, have an enormous potential derived from the sizeable amount of residues that could be used for energy generation (INPA 2011). The Federal State of Pará (PA) is the greatest wood producer in the country, accounting for 53.9% of the national wood production (Rendeiro, 2010). Many of the sawmills (65%) located in Pará process 6,000 to 12,000 m³ of timber per year, generating approximately 3,000 to 6,000 m³ of residue per year. More than 70% of the enterprises could use the residue and generate 200 to 300 kW of electricity (Rendeiro, 2010).

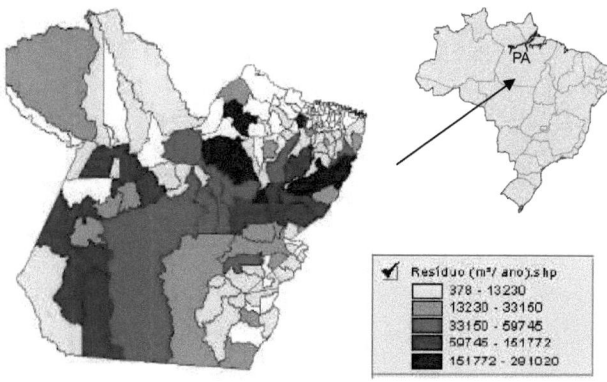

Figure 2.4: Wood residues available (m³ / year) in Pará (PA) (Rendeiro, 2010)

The wood residues are mostly used for cooking and charcoal production. The unrestricted disposal of sawdust is also very common in the Amazon region. The high quantity of residues can negatively affect the regeneration and growth of the native forest. The residues also increase the risk of forest burning and hinder forest growth. The great majority of the sawmills eliminate the residues by open-air combustion. Some industries integrate wood milling to charcoal production or generate steam for drying the saw wood (Rendeiro, 2010).

Figure 2.5: Wood residues disposed in a football field in Pará (own source)

Since the 1990s, timber exploitation is monitored and restricted to sawmills, which have environmental permition. In 1992, the Brazilian forestry sector developed a certification program that would identify the origin of raw materials used by the forest industry of Brazil. This work resulted in CERFLOR (Certificate of Origin of Forest Raw Material) (MDIC 2011, Inter-American Development Bank 1999).

According to IBGE (2010), between 1990 until 2007, the level of deforestation diminished in the whole country. However, illegal timber falling still occurs and huge areas of the Amazon forest have been transformed into pasture and soy monoculture. In the Amazon region, the markets are flooded with cheap timber from illegal operations (Nasi et al. 2011, De Lima et al. 2008). Official data regarding the amount of saw dust and wood leftovers from log processing in the Amazon region are inaccessible.

Soy expansion into the Amazon began in the late 1990s due to the growing demand for soy beans combined with low land prices and improved transportation infrastructure of south-eastern Amazon. Indirect deforestation was also caused by extended cultivation of soy on pasture land for cattle breeding in the transition area between the Cerrado and the Amazon (Nepstad et al. 2006).

2.2.4.2 Short-rotation planted forests

The total forest area in Brazil reached approximately 6.2 millions of hectares in 2009, being 4.3 million of hectares covered with eucalyptus trees and 1.9 million hectares with pine trees (ABRAF, 2009). The average annual yield for eucalyptus trees is 25 to 50 t/ha during a period of 7 years. For pine trees, the average yield is 20 to 40 t/ha during a period of 12 years.

According to IBGE (2010), short-rotation forests are used in several sectors: pulp and paper production (38%), timber production (28%), firewood for cooking (24%) and charcoal (10%). Mostly pine trees are used in the timber industry.

The Brazilian Institute for Forestry Development (IBDF), founded in 1967, created the first programs for reforestation. During the 1980s, the cultivation of eucalyptus and pine trees was supported by the government and there was an increase in charcoal production. This policy was criticized regarding its environmental impacts, because most of the charcoal was based on native forests. Therefore, new regulations fostered the production of charcoal using planted forests.

The paper and pig iron industry are planting short-rotation trees for energetic purposes. The enterprise PLANTAR was founded in 1967 and is today one of the greatest eucalyptus producers for charcoal production. Since 1985, it produces pig iron using charcoal (Grupo Plantar, 2011a). The eucalyptus forests are certified with the international criteria from the Forest Stewardship Council (FSC). During the forest growth, the trees capture CO_2 from the atmosphere and liberate oxygen. During the realization of a carbon trading project, the enterprise planted 23 thousand hectares of eucalyptus. The pig iron produced with eucalyptus charcoal avoided the emission of 13 million tons of CO_2. The project has duration of 28 years and should create more than a thousand jobs (Grupo Plantar, 2011b).

Brazil is a tropical country and the demand for household heating systems is limited. However, wood logs and wood residues from native and planted forests are widely used in rural areas for cooking. Biomass is burned in ovens with low efficiency and high emission of pollutants (Figure 2.6). The energetic efficiency of the traditional firing is low and the health risks are considered high.

Figure 2.6: Small-scale use of biomass: firewood stove (Fogão a Lenha: Fotos e Modelos, 2010)

According to Ludwig et al. (2003), there is an elevated indoor pollution level resulting from cooking. Trace gases and aerosols released through biomass burning have a significant influence on health, climate and biogeochemical cycles.

Much of the firewood is obtained by individual gathering and so the fuel use is not available. Such useful information would include: the consumption of firewood and the emissions per unit quantity of burned fuel.

The forestry sector is responsible for 0.5 million direct jobs and 2.5 million indirect ones. The largest companies, which cultivate eucalyptus or pine trees, are implementing an "integrated system" which consists of a partnership between industry and landowners, mainly small-scale farmers. Forestry industries like KLABIN in South Brazil provide seedlings and all other agricultural materials. Farmers participate with land and work force. Over 17 municipalities in the state of Paraná have been supported with inputs and technical assistance within the "Forestry Incentive Program" (KLABIN, 2011).

Wood processing generates a high amount of residue. The generation of waste is significant in the regions that exploit wood resources. In Brazil, pulp and paper factories face problems of waste disposal (approximately 48 tons of wood residues to 100 tons of pulp produced). The landfill option for final disposal is not feasible, due to the high costs of implementation and maintenance (Bellote et al. 1998).

The compaction of wood residues such as sawdust does not have a great expression in Brazil yet. In São Paulo, there are approximately 5 thousand pizza-houses and 8 thousand bakeries, in which 70% of energy comes from charcoal or firewood. This is gradually being changed to wood briquettes (Walter and Dolzan 2009).

Briquettes or pellets are more uniform, cleaner, and have a higher density than sawdust, which improves storage, transportation and combustion. The Brazilian market for these types of solid biofuels is young. The author visited one enterprise, which started producing pellets in Southeast Brazil, hoping to export than to Holland.

2.3 The Brazilian energy context

Hydropower accounts for 15.3% of the internal energy supply (Figure 2.7) but more than 76% in the electricity sector. In Brazil, about 13.9% of the energy needs are supplied by biomass (EPE, 2010). Sugar cane bagasse, charcoal and firewood are mainly consumed in the ethanol and pig iron industries, and in the residential sector respectively.

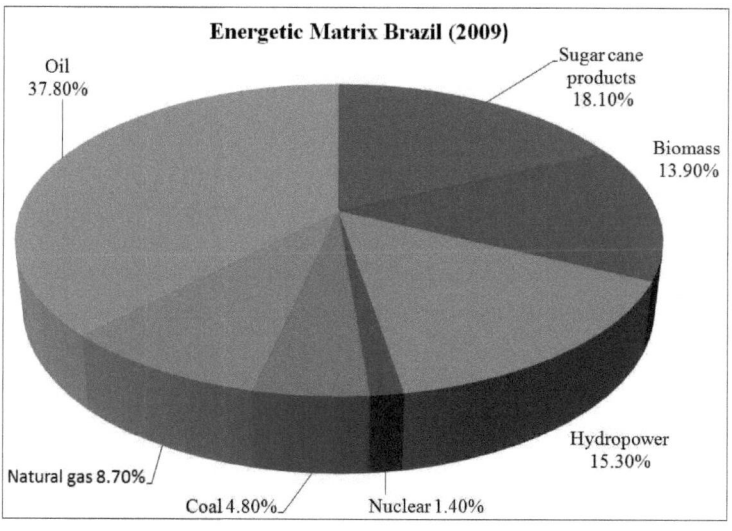

Figure 2.7: Internal energy supply in 2009 (EPE, 2010)

Firewood use is mostly residential and its consumption in rural areas has halved since the 1970s (Figure 2.8). The main reason is the introduction of gas stoves, substituting for traditional firewood cooking in rural areas. It is important to note that the household wood consumption is based on estimations (Walter and Dolzan, 2009). The geographical dimensions of the country hinder the collection of accurate data regarding biomass consumption.

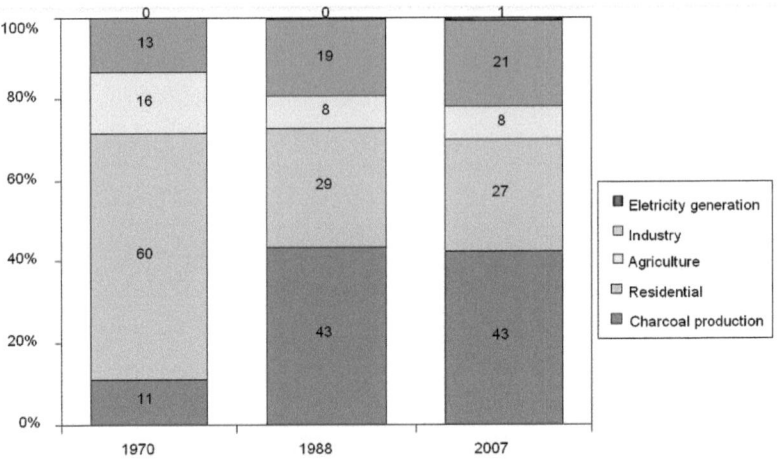

Figure 2.8: Wood consumption in Brazil (MME, 2007)

There are positive and negative consequences arising from this transition. For instance, environmental degradation through deforestation could be diminished. Furthermore, modern cooking technologies cut back hazardous emissions. In 2007 almost 40% of firewood was still consumed in Brazilian home kitchens (Figure 2.8).

Charcoal is mostly used in industry, specifically in the metallurgic industry. In the last few years, the Brazilian government implemented policies aiming at substituting coal coke for charcoal in the iron and steel production (Walter and Dolzan, 2009), thereby increasing charcoal production. However, due to the ups and downs of coal coke prices, charcoal use for pig-iron production is concentrated mostly in small-scale factories.

There is a growth tendency in the electric energy consumption as shows Table 2.3.

Table 2.3: Electric energy consumption supplied by the grid (GWh) (EPE, 2011)

Region	2009	2010
North	23,958	25,514

Northeast	64,596	70,096
Southeast	206,567	222,531
South	66,204	70,322
Center-West	24,699	25,901
Total	386,024	414,364

In 2008, new conditions of access to the national grid had been defined by the ANEEL opening prospects for the connection of ethanol and biomass plants to the grids (ANEEL, 2008). The evolution of regulations, legislations and official programs also stimulates the undertaking of renewable energy projects. Hence, to ensure the supply of electricity is necessary a diversification of primary energy generation, prioritizing technologies that use the available energy resources at a viable cost.

2.3.1 The power sector

The electricity industry in Brazil has undergone a series of significant institutional changes over the last century. Private ownership was not allowed from 1935 until the era of reforms of the 1990's. International trends in the electricity sector towards reform and the disproportion between investment costs and revenues led to a period of change. Some of the distribution utilities were privatized, independent power producers were encouraged, and the independent regulator ANEEL was established. While some utilities were privatized, this did not change the highly centralized nature of the Brazilian electricity system. Much like in the USA, the system is based on regional monopolies granted on a concession basis. Electricity providers have exclusivity in their region and universal supply and regulated tariffs (Zerriffi, 2008).

The installed capacity of Brazil in 2009 considering all the existing electrical sites and the international connections was 106 MW.

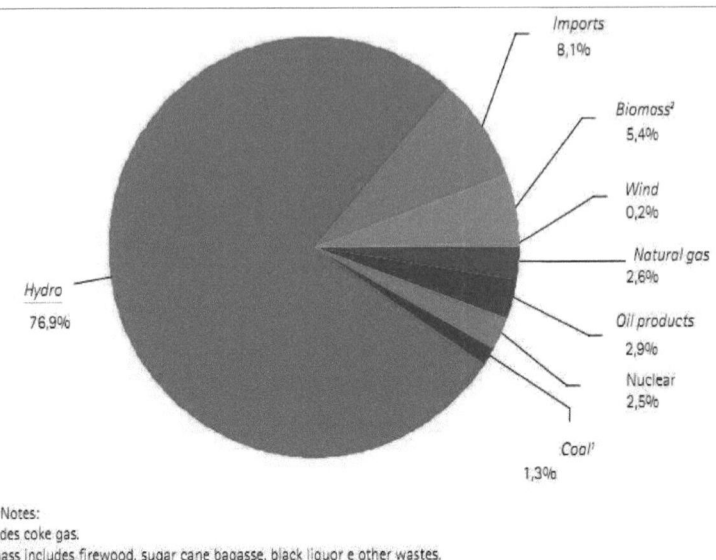

Figure 2.9: Domestic electricity supply in 2009 (EPE, 2010)

From this amount, biomass counted with 5 MW as shown in Figure 2.9 (EPE 2010). In the past, hydropower represented more than 90% of the installed power capacity in Brazil. This contribution has been changed gradually and in 2008 hydropower accounted for 74% of the total electricity supply in Brazil. The reason was the construction of new power plants supplied by biomass and natural gas.

As shown in Figure 2.10, the large-scale biomass power plants are currently located mainly in the Federal States of São Paulo (SP), Minas Gerais (MG), Espírito Santo (ES), and Goiânia (GO), which are the largest producers of rice, sugar cane and eucalyptus wood, respectively. The residues originated from these crops supply the biomass power plants.

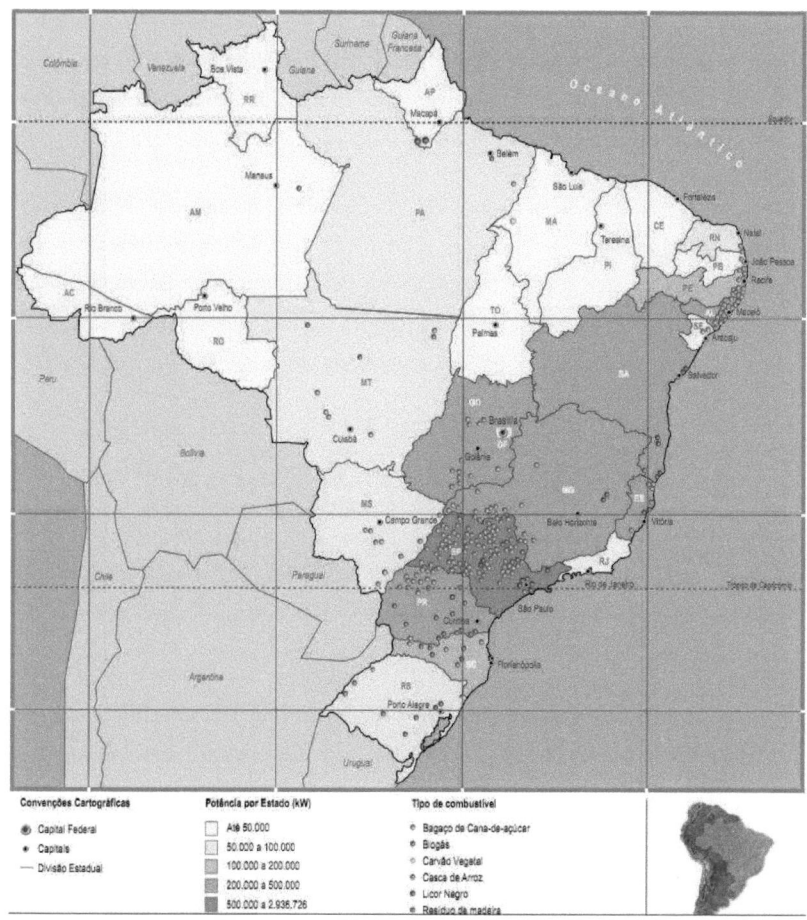

Figure 2.10: Biomass power plants installed capacity in each state and main crop locations (ANEEL, 2008)

According to ANEEL (2008) there are 302 biomass-fired power plants in Brazil, delivering 5,700 MW. Of the total plants listed, 13 are supplied with black liquor (944 MW), 27 per timber (232 MW), 3 by biogas (45 MW), 4 by rice husks (21 MW) and 252 for sugar cane bagasse (4,000 MW), One of the characteristics of these plants is their small installed capacity of 60 MW, favouring installation close to supply and demand centers.

Electricity from bagasse provides already 3% of the total Brazilian electricity needs, a figure expected to increase to 14% by 2020. This growth is possible as projections calculate that sugar cane production is going to increase by 340 million tonnes in the years to come, due to investments in 90 new sugar cane plants (Desplechin 2011).

Sugar cane has a great potential for electricity generation through use of bagasse and straw. This is important not only for diversification of the energy sector, but also because the sugar cane harvest coincides with the period of dry weather, where the largest installed capacity of hydroelectric plants is located. The electricity supplied by sugar cane bagasse in this period helps, therefore, to keep the reservoir levels for hydropower.

2.3.2 Electricity distribution and transmission in Brazil

Mostly large government-controlled companies, but also several private power producers, sell their electricity to distributors via auction. Electricity distribution is mainly operated by the private sector, whereas transmission is both publicly and privately owned.

In Brazil, the transmission lines vary from 88 kV to 750 kV. When the electricity arrives at the substations through a system composed of wires and transformers, these values are decreased to 127 volts or 220 volts (ANEEL, 2008).

The provision of electricity relies on government concessions, and the country is fully covered by either private (national or international) or state-owned (municipal, state or federal) concession areas (OECD/IEA, 2010). Currently there are 63 electricity providers, responsible for distributing electricity (ANEEL, 2008). These are large-scale enterprises, which have the monopoly of electricity in a region. In most federal states, only one provider operates. However, a state can have more than one provider, for example São Paulo and Minas Gerais. Prices can vary from BRL 0.19/kWh in North Brazil to BRL 0.43/kWh in Minas Gerais (ANEEL, 2008).

2.3.3 The Brazilian grid systems

Due to the large distance between generating sources and power grids, the electricity system is predominantly based on sites with central hydroelectric plants and a complex mesh of transmission and distribution. Hydropower generation has the largest contribution in the electricity system and an already established structure. In rural areas of Southeast Brazil, decentralized energy generation using biomass in small-scale steam cycles is not implemented due to the low electricity prices of hydropower generation (ANEEL, 2011).

Figure 2.11: The National Grid (ANEEL, 2008)

The Brazilian power sector is divided into two large systems, the interlinked (National Grid) and the isolated (Goldemberg, 2002). The National Grid, with approximately 80,000 MW of installed power, includes the Northeast-Southeast-South transmission line (Figure 2.11).

The isolated system includes small local grids mainly in the Amazon (Goldemberg et al., 2003). These 250 isolated systems are mainly thermal power generation from biomass and diesel. They are powered by 1230 generator units, supplying 3% of the Brazilian population, located in 380 localities, and represent 45% of the total territory (ELETROBRAS, 2009). Due to thermal energy conversion based mostly on diesel, the isolated systems have generation costs higher than that of the National Grid (Figure 2.12).

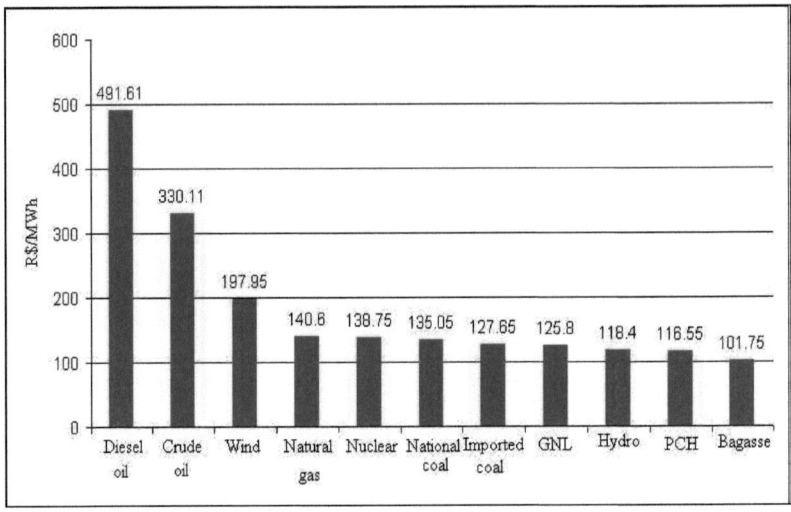

Figure 2.12: Production costs of electricity in Brazil based on different raw materials (ANEEL, 2008)

The difficulties in supplying the Amazon region with diesel increase the freight prices of the fuel and further the generation costs. Therefore, the government created the subsidy CCC[3], which is used for supporting energy generation in isolated regions using diesel. All electricity consumers in Brazil have this subsidy included in their electricity bills. In 2008, the CCC subsidy was of BRL 3 billion.

2.3.4 Electricity demand and supply

Brazil's residential sector of 42.5 million households consumes 25.2% of all electricity (Pereira et al. 2010). 97 % of all Brazilian households have access to regular electric energy. Whereas Brazil's urban sector accounts to almost 100% of service, in rural areas only 97% of households have access to electricity.

The North has the greatest levels of lack of electrification, with 62% of the rural population (about 2.6 million people) having no access to electricity. In Northeast Brazil, 39% of rural residents (about 5.8 million people) lack electricity. In the Center-

---
[3] Conta de Consumo de Combustíveis Fósseis = Subsidy for fossil fuel consumption

West the percentage is 27% (about 370,000 persons), in the Southeast 12% (about 810,000), and in the South 8.2% (about 480 000 people) (MME, 2003).

Urban areas are well electrified, but in the rural areas, 20 million of Brazilians have no access to electricity (MME, 2003). The electricity supply in the North is based on diesel generators. Overall fuel consumption is high not only due to the electricity generation itself, but also to local transportation, which is mostly by motorboat. There are presently ca. 1,000 power plants supplying electricity to isolated cities and villages in the Amazon using diesel oil. Almost 700 units have an installed power capacity below 500 kW (Goldemberg et al., 2003).

The electrification of isolated communities using diesel still represents a significant barrier to community residents (Goldemberg et al., 2003). Increasing electricity supply plays an important role in the increase of living standards because it allows high quality lighting, clean water, health care, and communications (Goldemberg et al., 2003). The lowest electricity consumption levels have been found in the households with low income. Within this framework, the Brazilian government has already pursued policies to promote the development of decentralized energy systems to provide energy to rural communities. The most important electrification programs are described in the next chapter.

2.3.5 Electrification programs

In recent years, government programs have linked millions of Brazilians to electricity through grid extensions. The **Program of Energy Development of States and Cities (PRODEEM)** was implemented by the Ministry of Mines and Energy (MME) in 1994 as an initiative to bring electricity to rural communities using renewable resources, mainly solar energy. The government supplied communities with equipment and distributed it to the chosen projects. The projects were focused on community development rather than household electrification (WRI, 2011).

Between 1996 and 2000, PRODEEM provided 3 MW power to 3,050 villages, benefiting 604,000 people. The National Treasury Fund supported the program with BRL 21 million, investing in photovoltaic panels for rural power generation (EIA, 2011). According to Goldemberg et.al (2003), only 44 out of the 79 solar systems (56%)

were actually operating in 43 villages located in 10 States. Problems with PRODEEM include the following:

- A top-down approach, with installations sometimes made in unskilled and unorganized communities,
- No schemes for cost recovery, which results in a lack of funds for maintenance and hence unsustainable service,
- Lack of responsibility of local communities and federal states for the equipment
- Occasional lack of coordination with grid expansion programs and difficulties in identifying suitable locations for equipment purchased in bulk.

In 1999 the **Program "Luz no Campo" (Light in the Countryside)** was implemented by ELETROBRÁS under the coordination of the MME. It intended to electrify one million rural homes in a three-year period. Rural consumers paid the full costs of the connection, spread out over a number of years. This program lent 75% of the investment to concessionaires at 6% interest, with a two-year grace period and a five to ten year repayment period (WRI, 2011). By 2002, this program had connected 480,000 households and another 125,000 were in progress.

The **Program "Luz para Todos" (Light for All)** was implemented in 2003 and aimed to provide social inclusion and reduce poverty and hunger through electricity supply to 12 million inhabitants in 5 years (OECD/IEA, 2010). In this program, tariffs were to be reduced, energy to be provided free to low-income consumers and residential consumers with consumption less than 80 kW/month (WRI, 2011). The projected costs were BRL 14 billion. This sum was achieved through a partnership of the federal government, state agencies and energy distributors. The federal government designated BRL 10 billion. The project is also supported by US AID funding (WRI, 2011). Due to the high need of electricity connections, the program was postponed until the end of 2011 (MME, 2011 a).

The coverage rate in rural areas has increased from 49% in 1991 to 97% in 2008 (Pereira et al. 2010). The positive result can be partially explained by the reduction of the rural population in the country (around 5 million), and together with the greater effort made by the program Luz para Todos. However, this program has no clear strategies to support and educate isolated communities regarding the use of local energy in an efficient way.

2.3.6 Policies and measures

The MME oversees the whole power sector and is responsible for policy setting. The Brazilian Electricity Regulatory Agency (ANEEL), operating under the jurisdiction of the MME, regulates and controls the generation, transmission and distribution of power in compliance with current legislation (OECD/IEA, 2010).

ELETROBRÁS is a state-owned holding company for electricity assets, controlling a large part of electric power generation and transmission systems mainly through subsidiary electricity providers, as well as some distribution capacity in the Amazon area. The Energy Electricity Research Agency (EPE) is responsible for the development of integrated long-term planning of Brazil's power sector, supporting MME's national energy policy and long-term planning.

The electricity subsidy is distributed to the electricity providers, which have exclusive service and do not allow the consumer to make decisions based upon their energy needs and the availability of different options for meeting those needs. The generation of electricity in Amazonas is controlled by government-owned centralized utilities like the CELPA in the Federal State of Pará. It serves the rural areas with electricity mostly based on diesel generation. The capacity of grids supplied by diesel allows low quality household electrification and only few hours per day. The cost of diesel electricity is high. As official electricity providers, CELPA is able to utilize the "Luz para Todos" funds plus the subsidy for diesel to keep costs down.

The diesel subsidy[4] was created in 1973 and is still in force today. It is managed by ELETROBRÁS, which gives quotas to the electricity providers, which are proportional to the amount of trade energy (Andrade et al., 2011). The budget variations is based on hydrologic forecasts, the expansion of the market, fluctuations in the price of fuel and the specific consumption of generating centers, leading to greater or lesser fuel consumption, with a direct impact on generation costs.

In 1993 the Federal Government significantly altered the structure of the electricity sector, with the aim of stimulating competition and attracting private-sector investors. After the reform, electrification in rural areas have progressed from 68% to 74% of househoilds (Goldemberg et al., 2003). In 2002 the Law 10438 established rules and

---

[4] Conta de Consumo de Combustíveis (CCC)

obligations of distribution concessionaires and introduced changes into the structure of Brazilian energy sector. According to the MME (2002) the following measures are the most important:

- Type of consumers: low-income consumer (80 kWh / month consumption), and high-income consumer (200 kWh / month) and
- Subsidies to renewable energy sources through the Program of Incentives for Alternative Electricity Sources (PROINFA).

More than a third of residential customers (35%) in Brazil belong to the low consumer tariff class. A total of 77% of the low-income consumers uses up to 80 kWh of electricity per month and 23% between 80 and 220 kWh (Instituto Acende Brasil, 2007). Electricity use for the rural poor is promoted essentially through lower tariffs for low-consumption classes, the idea being that low consumption is correlated with low income (Zerriffi, 2007).

Consumers, who use up to 30 kWh/month, pay only 35% of the regular residential tariff, while those consuming between 31 kWh/month and 80 kWh/month pay the reduced tariff of 60% of the regular residential tariff.

In the states of the North and the Northeast, 43% and 62% of all consumers respectively fall within the subsidized tariff classes, which are the highest rates for all Brazil (Instituto Acende Brasil, 2007). It should be noted that these regions also display the highest poverty indices and lowest access rates to a variety of public services. Although low income does not always correlate with low consumption levels, recently connected households that live below or near the poverty line rarely consume high amounts of electricity. According to ANEEL (2011) in 2010-2011, the average electricity tariff in Brazil was BRL 0.32 / kWh (USD 0.19 / kW/h).

The PROINFA became the main program to promote electricity generation from renewable sources through incentives and subsidies. It was designed to achieve 10% of Brazilian power production and this goal is supposed to be reached within the next 20 years (MME 2008). In the first phase, PROINFA required the installation of 3,300 MW from wind turbine, small hydropower stations and biomass plants. Long-term purchase guarantees were offered to independent power generators: produced electricity is to be purchased by ELETROBRÁS at fixed feed-in prices for 20 years. The fixed purchase prices vary amongst renewable sources and are determined by the economic value of the competing energy source.

The Brazilian National Development Bank (BNDES) makes special financing programmes available for renewable projects that are eligible for PROINFA. The BNDES can finance up to 70% of capital costs (excluding site acquisition and imported goods and services) at the basic national interest rate. Interest is not charged during construction and amortization is of 10 years. Payments are due 6 months after commercial operation (EIA, 2011).

Interest rates in Brazil are regarded to be among the highest ones in the world. Therefore, long-term business for power generation become unattractive under rates of 20 to 30%/year. The BNDES has lower interest rates, ranging from 10-18%/year. Another issue is land prices in Brazil, which are relatively low compared to land prices in Europe. This helps mitigate higher financial costs (Walter and Dolzan, 2009).

PROINFA is expected to generate 150,000 jobs and to leverage private investments of ca. BRL 5 billion. The legislation mandates the use of at least 60% national products and services from total construction costs. Each federal state has maximum limits of 20% of total capacity for wind and biomass and 15% for small hydropower plants. Such limits are preliminary: if part of the 1,100 MW for each source is not purchased, this available potential will be distributed according to the older environmental permits. All projects are required to have installation permission (EIA, 2011).

By early 2005 the first phase was finished and 3,300 MW was installed (1,266 MW Solar, 655 MW Biomass and 1,379 Wind). In 2010, 68 plants were operating: 23 small hydropower plants (414.30 MW), 2 biomass plants (66.50 MW) and 43 wind mills (1,110.97 MW), adding a total of 1,591.77 MW to the national grid (MME, 2011 b).

2.3.7 Auction schemes for power generation projects

The Brazilian power sector model holds energy auctions aiming supporting new power generation projects. The main objectives of the auctions is to guarantee that the offer will expand in such a way as to satisfy the growth of demand with the lowest possible generation cost in line with the need of guaranteeing low tariff (Castro and Dantas 2010). This means, the winning projects are those providing the lowest electricity tariff.

For example, the auction of the hydropower plant Madeira River (ANEEL, 2008) sold 70% of produced electricity to the electricity providers at a lower price (BRL 0.78 / kWh) compared to the price established by the MME (BRL 1.22 / kWh).

The ANEEL classifies power plants into three categories, those without restrictions, moderate restrictions or serious restrictions. The plants without restrictions have signed the concession contracts, have a valid environmental license and are following the construction schedules (Castro, 2007).

As an example, the Reserve Energy Auction carried out in 2008, specific for advancing bioelectricity, has reached 3,518 MW, representing 24% of the total power projects (UNICA, 2010). The licensed biomass projects are located in 9 Federal States: São Paulo is the leader with 32 projects (1,873 MW). Mato Grosso do Sul has 7 projects (596 MW) and Minas Gerais has 6 projects (537 MW). The biomass projects have a duration of 15 years.

The 2008 auction has commercialized energy at an average cost of BRL 155.70/MWh. As a comparison, the generic price was BRL 145.23/MWh. According to Castro et.al (2008), sugar cane projects would be viable at a tariff of BRL 155/MWh without commercializing carbon credits. Therefore, bioelectricity projects are on threshold of competition with conventional thermal projects.

2.4 Discussion

Brazilian crops and their harvesting and processing residues hold great potential for energy generation. As an example, bagasse co-generation can be considered as a win-win for the sugar industry located in São Paulo (WADE, 2004), because it combines low cost, efficiency and social benefits with the provision of clean and renewable energy. Another successful example is the cultivation of eucalyptus trees for supplying the iron ore production and the paper industry. The wood residues are not fully used.

The creation of incentives for increasing the generation and distribution of electricity in Brazil during the 1990s was essential. These measures allowed biomass power plants to have better access to financing. Furthermore, the energy commercialization costs were diminished and the energy purchase and sale contracts could apply over longer periods.

In addition, the Clean Development Mechanism (CDM), included in the Kyoto Protocol, was an important policy. The CDM stimulated Brazilian and foreign investment in biomass projects to obtain additional income by selling carbon credits to polluting companies, which must reduce the GHG emission levels.

Investors in biomass projects are still exposed to the risk of not fulfilling PROINFA's commitments, not receiving financing, facing long environmental licensing process for each power plant, and also obtaining the connection point to the distribution grid (Castro, 2007).

According to Zerriffi (2008), Brazil's official rural electrification system has three main characteristics as a result of the regulatory regime and policies of the central government:

1. Exclusive service territories for the electricity providers with a requirement for electricity providers to provide electricity to all consumers within the territory
2. Electricity subsidies
3. Low tariffs for low income and rural costumers

Since all the customers are in the rural and lower income group, the electricity providers also rely on its status as a government utility, part of the ELETROBRÁS group, which allows it to run a deficit. The losses in 2004 were BRL 71 million (Zerriffi, 2008). This is a direct result of the regulations governing the electricity sector, which mandate exclusive service territories for electricity providers and low tariffs for low-income consumers.

Rural electrification initiatives such as the program "Luz para Todos" has concentrated mostly on increasing access to infrastructure. There are still ten million people in Brazil without electricity. Access to the grid is still a barrier to new biomass projects, which are located in remote areas. Many of them are distant from the energy substations, which are the distributors of the produced electricity.

Electrification projects based mainly in grid connections are not generally economically viable (Goldemberg et al., 2003). Energy supply in these cases must be decentralized, and there is an excellent opportunity for the introduction of renewable energy. Community participation in electricity management is fundamental in remote areas to reduce costs. The social and economic benefits must be maximized to rural communities through the implementation of productive activities (Goldemberg et al., 2003).

Access to electricity is essential to economic development as is indispensable for basic activities, such as lighting, refrigeration and the running of household appliances. Brazil has a vast potential in the utilization of agricultural and wood residues for solid biofuel production for heat and electricity generation in decentralized plants. In this regard, technical and economical improvements could be achieved by industrial upgrading (for instance by biomass compaction). The pretreatment of these raw materials could expand the perspectives of the use of biomass. The next chapter gives an overview of the state of the art of biomass technologies, which could be applied in Brazil.

2.5 References

ABRAF - Associação Brasileira de Produtores de Florestas Plantadas (2009) Available at : http://www.abraflor.org.br/ (Last access 08 June2011)

Andrade CA., Rosa LP. and Silva NF. (2011) Generation of electric energy in isolated rural communities in the Amazon region: a proposal for the autonomy and sustainability of the local populations. In: Renewable and Sustainable Energy Reviews 15: 493-503. Elsevier

ANEEL – Agência Nacional de Energia Elétrica (2008) Atlas de energia elétrica Brasil. 3a. edicao. Available at: http://www.aneel.gov.br/arquivos/PDF/atlas_par1_cap1.pdf (Last access 16 May 2011)

ANEEL - Agência Nacional de Energia Elétrica (2009) Available at: http://www.aneel.gov.br/?idiomaAtual=1 (Last access 16 May 2011)

ANEEL - Agência Nacional de Energia Elétrica (2011) Tarifas residenciais. Available at: http://www.aneel.gov.br/area.cfm?idArea=493&idPerfil=4 (Last access 31 May 2011)

ARE - Alliance for Rural Electrification (2010) Available at: http://www.ruralelec.org/ (Last access 8 June 2011)

Bellote A., da Silva H., Ferriera C. and Andrade G. (1998) Resíduos da indústria de celulose em plantios florestais. Boletim de Pesquisa Florestal, Colombo, n. 37, p. 99-106, Jul./Dez.

BNDES, CGEE, CEPAL, FAO (2008) Sugar cane based bioethanol: Energy for sustainable development. Available at: http://www.bioetanoldecana.org/ (Last access 8 June 2011)

Bourne J. (2007) Growing fuel: The wrong way, the right way. Official Journal of the National Geographic Society. Vol. 212. No.4

Briquet M. (2007) Brazil's ethanol program: an insider's view. Energy Tribune. Available at: http://www.energytribune.com/articles.cfm?aid=534 (Last Access 16 June 2010)

Castro M. (2007) The expansion of distributed generation in Brazil: analysis of the current incentives and the risks for investors. Available at: http://www.gwu.edu/~ibi/minerva/Fall2007/Marco.pdf (Last access 8 June 2011)

Castro N. and Dantas G. (2010) Planning of the Brazilian Electricity Sector and the World Context of Climate Change Available at : http://www.nuca.ie.ufrj.br/gesel/artigos/100525planningbrazilian.pdf (Last access 6 April 2011)

Clear (2010) Camil Itaqui Biomass Electricity Project. Available at: http://www.clear-offset.com/carbon-offset-projects.php (Last access 8 June 2011)

CPTEC - Centro de Previsão de Tempo e Estudos Climáticos (2011) Climatologia de temperatura e precipitação. Available at: http://www.cptec.inpe.br/ (Last access 6 Mai 2011)

Coelho S. (2009) Energy access in Brazil. Taller Latinoamericano y del Caribe: Pobreza y el Acceso a la Energía. CENBIO – Brazilian Reference Center on Biomass University of São Paulo, Santiago. Available at:
http://www.eclac.org/drni/noticias/noticias/6/37496/Coelho.pdf (Last access 6 April 2011)

Couto LC., Couto L., Watzlawick LF. and Camara D. (2004) Vias de Valorização Energética da Biomassa. In: Biomassa e Energia 1(1): 71-92 Available at: http://www.renabio.org.br/arquivos/p_vias_biomassa_5919.pdf (Last access 6 April 2011)

Department of State (2011) Backgroud and note: Federative Republic of Brazil. Secretary for Public Diplomacy and Public Affairs. Bureau of Public Affairs: Electronic Information and Publications. Available at: http://www.state.gov/r/pa/ei/bgn/35640.htm (Last access 6 Mai 2011)

De Lima A., Keppe A., Alves M., Maule R. and Sparovek G. (2008) Impact of FSC Forest Certification on Agroextractive Communities of the State of Acre, Brazil; Imaflora, University of São Paulo, Entropix Engineering Co.: São Paulo, Brazil.

Desplechin E. (2011) Brazilian bioelectricity from bagasse. Available at : http://english.unica.com.br/opiniao/show.asp?msgCode={73BF1846-A8CC-4494-943C-FC2BAE4B1C06}(Last access 6 April 2011)

EIA - International Energy Agency (2011) Global renewable Nergy – Poliy and measures. Available at: http://www.iea.org/Textbase/pm/?mode=re&id=1476&action=detail (Last access 31 March 2011)

ELETROBRÁS (2009) GTON- Grupo Técnico Operacional da Região Norte. Available at: http://www.eletrobras.gov.br/EM_Atuacao_SistIsolados/default.asp (Last access 11 May 2011)

EMBRAPA (2010) Seminário sobre briquetes de casca de arroz apresentou alternativa energética sustentável. Pecuária Sul, Breno Lobato (MTb 9417 - MG). Avaible at: http://www.embrapa.br/imprensa/noticias/2010/novembro/1a-semana/seminario-sobre-briquetes-de-casca-de-arroz-apresentou-alternativa-energetica-sustentavel/ (Last access 31 March 2011)

EPE - Empresa de Pesquisa Energética (2010) Brazilian Energy Balance 2010 (year 2009). Available at: https://ben.epe.gov.br/downloads/Relatorio_Final_BEN_2010.pdf (Last access 1 May 2011)

EPE - Empresa de Pesquisa Energética (2011) Série estatísticas energéticas. Nota técnica dea 05/11. Boletim de conjuntura energética. 3° trimestre 2010. Rio de Janeiro, Maio de 2011. Available at: http://www.epe.gov.br/mercado/Documents/S%C3%A9rie%20Estudos%20de%20Energia/20110505_1.pdf (Last access 17 May 2011)

Eriksson S. and Prior M. (1990) The briquetting of agricultural wastes for fuel. Food and Agriculture Organization of the United Nations, Rome, Italy. Available at: http://www.fao.org/docrep/T0275E/T0275E08.htm (Last accessed 12 Februar 2009)

ESMAP (2005) Brazil background study for a national electrification strategy – aiming for universal access. World Bank, Washington D.C.

FAO - Food and Agriculture Organization of the United Nations (1990) The briquetting of agricultural wastes for fuel. Publications Division. M-09, ISBN 92-5-102918-0. Available at: http://www.fao.org/docrep/T0275E/T0275E07.htm (Last access 13 April 2011)

Filho P. and Badr O. (2004) Biomass resources for energy in North-Eastern Brazil. In: Applied Energy 77 (1): 51-67

Fogão a Lenha: Fotos e Modelos (2010) Available at: http://globodicas.com.br/fogao-a-lenha-fotos-modelos/ (Last access 8 June 2011).

GESEL - Grupo de Estudos do Setor Elétrico (2011) Available at: http://www.nuca.ie.ufrj.br/gesel/ (Last access 8 June 2011).

GOB (Government of Brazil) - Presidência da República Federativa do Brasil. (2011) Available at: http://www.brasil.gov.br/ (Last access 8 June 2011).

Goldemberg J. (2002) The Brazilian Energy Initiative - Support Report Presented at the World Summit for Sustainable Development, Johannesburg.

Goldemberg J., Rovere EL. and Coelho ST. (2003) Expanding access to electricity in Brazil. Available at: http://www.afrepren.org/project/gnesd/esdsi/brazil.pdf (Last access 31 March 2011)

Goldemberg (2007) Ethanol for a Sustainable Energy Future. In: Science 9, Vol. 315 no. 5813 pp. 808-810 DOI: 10.1126/science.1137013

Grupo Plantar (2011a) Plantar Siderúrgica S.A: siderurgia gusa-verde. Available at: http://ravel.plantar.com.br/portal/page/portal/plantar/siderurgia/a_empresa (Last access 31 March 2011)

Grupo Plantar (2011b) Projeto crédito de carbono. Available at: http://ravel.plantar.com.br/portal/page/portal/plantar/projeto_carbono/num_proj (Last access 31 March 2011)

Hassuani SJ., Leal MR. and Macedo IC. (2005) Biomass power generation: sugar cane bagasse and trash. In: PNUD / CTC 1st Edition Piracicaba, Brazil. Available at: www.ctcanavieira.com.br/(Last access 31 May 2010)

IBGE - Instituto Brasileiro de Geografia e Estatística (2006) Available at: http://www.ibge.gov.br/english/ (Last access 8 June 2011).

IBGE – Instituto Brasileiro de Geografia e Estatística (2010) Sistema IBGE de Recuperaçao Automática SIDRA. Available at: www.sidra.ibge.gov.br (Last access 31 March 2010)

IBGE - Instituto Brasileiro de Geografia e Estatística (2011) Available at: http://www.ibge.gov.br/english/ (Last access 8 June 2011).

IICA - Instituto Interamericano de Cooperação para a Agricultura (2011) Universalização de acesso e uso da energia elétrica no meio rural brasileiro: lições do Programa Luz para Todos. ISBN13: 978-92-9248-329-6.

INPA - Instituto Nacional de Pesquisas da Amazônia (2011) Available at: http://www.inpa.gov.br/ (Last access 13 April 2011)

Inter-American Development Bank (1999) Trade and Environmental Issues in Forest Production. Social Programs and Sustainable Development Department. Environment Division Working Paper. Available at:
http://ctrc.sice.oas.org/geograph/environment/simula.pdf (Last access 13 April 2011)

Instituto Acende Brasil (2007) Tarifa de baixa renda. In: Cadernos de política Tarifária Available at: http://www.acendebrasil.com.br/archives/files/estudos//Caderno_06_Baixa_Renda.pdf (Last access 1 April 2011)

Jank (2008) Available at: http://english.unica.com.br/ (Last access 8 June 2011).

KLABIN (2011) Available at: http://www.klabin.com.br/pt-br/home/default.aspx (Last access 8 June 2011).

Lora E., Andrade R. (2009) Biomass as energy source in Brazil, Renewable and Sustainable Energy Reviews, Volume 13, Issue 4, May 2009, Pages 777-788, ISSN 1364-0321, DOI: 10.1016/j.rser.2007.12.004. Available at: http://www.sciencedirect.com/science/article/pii/S1364032108000270 (Last access 6 June 2011)

Ludwig J., Marufu LT., Huber B., Andreae MO. and Helas G. (2003) Domestic combustion of biomass fuels in developing countries: a major source of atmospheric pollutants. Journal of Atmospheric Chemistry 44: 23-27, Netherlands

Marcon R., Zukowski J., Cavalcante J., Lopes I. (2004) Analiação de planta térmica com biomassa (briquete de casca de arroz): Caso real "Fazenda experimental do centro universitário luterano de palmas". Procedings Energia no meio rural, Campinas. Available at: <http://www.proceedings.scielo.br/scielo.php?script=sci_arttext&pid=msc0000000220 04000100008&lng=en&nrm=iso> (Last access 10 Mai 2011)

Martine G. and Garcia R. (1987) Os impactos sociais da modernização agrícola. São Paulo, Caetés.

Mazzafera P. (2002) Degradation of caffeine by microorganisms and potential use of decaffeinated coffee husk and pulp in animal feeding. In: Scientia Agricola 59 (4): 815-821

MDIC - Ministério do Desenvolvimento, Indústria e Comércio Exterior (2011) Cerflor: Certificação Florestal. Available at: http://www.inmetro.gov.br/qualidade/cerflor.asp (Last access 1 April 2011)

MME – Ministério de Minas e Energia (2002) Lei 10438. Available at: http://www.jusbrasil.com.br/legislacao/anotada/2534003/art-3-par-5-da-lei-10438-02 (Last access 31 March 2011)

MME – Ministério de Minas e Energia (2003) Available at : http://www.mme.gov.br/mme (Last access 8 June 2011).

MME – Ministério de Minas e Energia (2007) Plano Nacional de Energia 2030. Available at: http://www.forumdeenergia.com.br/nukleo/pub/pne_2030_documento_final.pdf (Last access 1 March 2011)

MME – Ministério de Minas e Energia (2008) Available at : http://www.mme.gov.br/mme (Last access 8 June 2011).

MME – Ministério de Minas e Energia (2011 a) Programa Luz para Todos. Available at: http://luzparatodos.mme.gov.br/luzparatodos/Asp/o_programa.asp (Last access 31 March 2011)

MME – Ministério de Minas e Energia (2011 b) PROINFA. Available at: http://www.mme.gov.br/programas/proinfa/ (Last access 31 March 2011)

Nass L., Pereira P. and Ellis D. (2007) Biofuels in Brazil: An Overview. In: Crop Science 47 (6): 2228-2237. Available at: http://ddr.nal.usda.gov/handle/10113/11628 (Last access 6 June 2011)

Nasi R., Putz F., Pacheco P., Wunder S. and Anta S. (2011) Sustainable Forest Management and Carbon in Tropical Latin America: The Case for REDD+. Forests 2011, 2, 200-217; doi:10.3390/f2010200. ISSN 1999-490

Nepstad D., Stickler C. and Almeida O. (2006) Globalisation of the Amazon soy and beef industries: opportunities for conservation. In: Conservation Biology 20 (6): 1595-1603

OECD/IEA (2010) Comparative study on rural electrification policies in emerging economies. Available at: http://www.iea.org/papers/2010/rural_elect.pdf (Last access 31 March 2011)

Pereira MG., Freitas MA. and Silva NF. (2010) Rural electrification and energy poverty: empirical evidences from Brazil. In: Renewable and Sustainable Energy Reviews 14: 1229-1240. Elsevier

Rendeiro G. (2010) Geração de energia elétrica em localidades isoladas na amazônia utilizando biomassa como recurso energético. Tese de doutorado. Programa de Pós-Graduação em Engenharia de Recursos Naturais da Amazônia - PRODERNA. Instituto de Tecnologia.Universidade Federal do Pará. Agosto 2011.

Ribando C. and Yacobucci B. (2007) Ethanol and other biofuels: potential for U.S.-Brazil energy cooperation. Congressional Research Service, Latin American Affairs Foreign Affairs, Defense, and Trade Division/ Environmental and Energy Policy, Resources, Science, and Industry Division. Available at: http://www.wilsoncenter.org/news/docs/CRS%20Report%20on%20US-Brazil%20potential%20cooperation%20on%20biofuels.pdf (Last access 13 April 2011)

Ribeiro D. (1997) O povo brasileiro: a formacao e o sentido do brasil. São Paulo, Ed. Companhia das Letras, 476 p.

Rosillo-Calle F., de Groot P., Hemstock SL. and Woods J. (2007) The biomass assessment handbook: Bioenergy for a sustainable environment. Earthscan, London

Schnepf D., Dohlman E. and Bolling C. (2001) Agriculture in Brazil and Argentina: Developments and Prospects for Major Field Crops. Market and Trade Economics Division, Economic Research Service, U.S. Department of Agriculture, Agriculture and Trade Report No. WRS013, 85 pp. http://www.ers.usda.gov/Publications/WRS013/ Accessed 12 Februar 2009 (Last access 31 March 2011)

Shikida P. (2010) The economics of ethanol production in Brazil: a path dependence approach. Available at: http://urpl.wisc.edu/people/marcouiller/publications/URPL%20Faculty%20Lecture/10P ery.pdf (Last access 15 June 2010)

Smeets E., Junginger M., Faaij A., Walter A., Dolzan P. and Turkenburg W. (2008) The sustainability of Brazilian ethanol: an assessment of the possibilities of certified production. In: Biomass and Bioenergy 32: 781-813.

UNICA - União da Indústria de Cana-de-açúcar (2008) Setor sucroenergetico - histórico: ciclo econômico da cana-de-açúcar. Available at: http://www.unica.com.br/content/show.asp?cntCode=8875C0EE-34FA-4649-A2E6-80160F1A4782 (Last Access 16 June 2010)

UNICA - União da Indústria de Cana-de-açúcar (2010) Setor sucroenergetico registra Available at: http://www.unica.com.br/noticias/show.asp?nwsCode={E52DBED8-BA81-4FDB-9EDA-37590AB18DDE} (Last Access 16 June 2010)

UNICA - União da Indústria de Cana-de-açúcar (2011) Available at: http://www.unica.com.br/ (Last access 8 June 2011)

WADE – World Alliance for Decentralized Energy (2004) Bagasse co-generation: global review and potential. Available at: http://cdm.unfccc.int/filestorage/2/K/J/2KJXDVUHFZ0MYG3WOBT91NA7QE6LP5/4057%20Annex%206%20Bagasse%20Co-generation%20-%20Global%20Review%20and%20Potential.pdf?t=cmF8MTMwNTU0MjkyMi40OQ==|SYSVFPJYJyez_gsY1e70e2aM_bs= (Last access 16 April 2011)

Walter A. and Dolzan P. (2009) Brazil country report – 2009. In: Country Report: Brazil – Task 40 – Sustainable Bio-energy Trade; securing Supply and Demand. Available at: http://www.bioenergytrade.org/downloads/brazilcountryreporttask40.pdf (Last access 16 April 2011)

Winfield E. (2008) Ethanol in Brazil. The University of Iowa Center for International Finance and Development. UICIFD Briefing No. 6. Available at: http://www.uiowa.edu/ifdebook/briefings/docs/brazil.shtml (Last access 16 June 2010)

WRI - World Resources Institute (2011) Special Programmes. Available at: http://projects.wri.org/sd-pams-database/brazil/national-programme-energy-development-states-and-municipalities-prodeem (Last access 6 June 2011)

Zerriffi H. (2008) From acaí to access: distributed electrification in rural Brazil. International Journal of Energy Sector Management 2, pp. 90-117

3 Energy from Biomass

Bioenergy is expected to contribute significantly in the mid and long term for future energy supply. Biomass currently supplies only a small fraction of the world's demand, although the contribution of the renewables seems likely to grow rapidly in coming decades. According to International Energy Agency (IEA), bio-energy offers the possibility to meet 50 % of our energy needs in the 21th century (Soetert and Vandamme 2009).

Biomass is the organic material derived from plants. Aquatic and terrestrial vegetation absorb sunlight, water and CO_2 to produce biomass through a process called photosynthesis. Plants are mainly constituted of carbon. When growing, the biomass first removes the greenhouse gas CO_2 from the atmosphere and binds the carbon in the biomass. This carbon is later released into the atmosphere again as a result of combustion or when the biomass is rotting. That means the biomass, used for energy purposes, releases only the CO_2 that was previously removed from the atmosphere when the plant was growing (Dürrschmidt, 2006). There is therefore a closed carbon cycle. According to Sims (2003), the term biomass includes:

- crop wastes (eg. cereal straw and maize straw for heat and electric generation);

- agro-industrial residues (eg. rice husks and bagasse for heat and electric generation)

- animal wastes (eg. sewage sludge to produce biogas)

- woodlot arisings (eg. from agroforestry and siviculture after log extraction and used mainly for heating);

- forest residues (eg. remainings after log extraction or wood process residues at the sawmill or pulp plant);

- municipal solid waste (either combusted in waste-to-energy plants or placed in landfills with the methane gas collected); and

- energy crops (eg. oil crops to produce biodiesel, or sugarcane and maize for bioethanol, or short rotation coppice for heat and electric generation).

Biofuel is a general term, which includes any solid, liquid, or gaseous fuels produced from organic matter, either directly from plants or from industrial, commercial, domestic or agricultural wastes. Bioenergy covers all energy forms derived from organic fuels (biofuel) of biological origin used for energy production. It comprises purpose-grown energy crops, as well as multipurpose plantations and by-products (residues and wastes). The term "by-products" includes solid, liquid and gaseous by-products derived from human activities (Rosillo-Calle, 2007).

According to the Food and Agriculture Organization (FAO, 2009), bioenergy provides about 10 percent of the world's total primary energy supply. Most of it is used in the residential sector for heating and cooking and is produced locally. In 2009, 97% of the biofuels used in the world were counted as solid biomass, 71% of which used in the residential sector. Wood fuels count to about 50% of the total combustible renewables and waste used in the world (FAO, 2009).

Solid biomass such wood and agricultural residues, short-rotation woody and herbaceous crops can be used for the production of heat and electricity in different processes, depending on the type of material available. The development of small-scale biomass power plants creates in all stages of crop production, a multitude of possible working positions and different trading schemes. This new situation could help changing the traditional energy consumption structures and preserving farmer's culture. However, the improvement of the current biomass technologies needs innovation in all energy generation scales.

Biomass is considered as a promising energy source able to contribute with the greenhouse gas emissions reduction targets. Biomass combustion is credited be CO_2-neutral and to releasing much less sulphur, methane, heavy metals and particulate emissions than their fossil fuel counterparts. The current global energy patterns, which are mostly based in fossil fuel consumption, could justify the increase in the use of biomass. The author gives an overview regarding global energy production and climate change in the following subchapter.

3.1 Global energy production and climate change

Nowadays, fossil fuels supply nearly 80% of the global current energy consumption (Figure 3.1). However, petroleum, coal and natural gas can be found in relatively few countries. This concentration has already led to economic disruption and political conflicts, such as the Oil Crisis in the 1970's and the Gulf War in the 1990's.

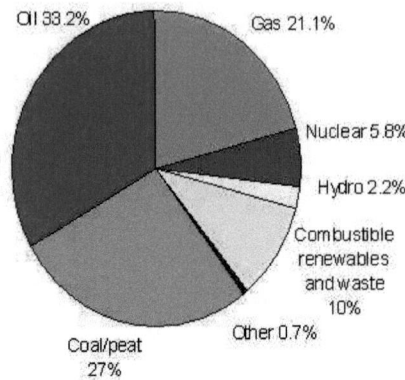

Figure 3.1: Fuel shares of total primary energy supply in 2008 (Source: IEA, 2010)

The massive use of coal, oil and gas has increased material prosperity in the majority of the industrialized countries. The primary energy demand has grown by more than 50% since 1980. Forecasts predict that this growing trend will continue during the next 50 years.

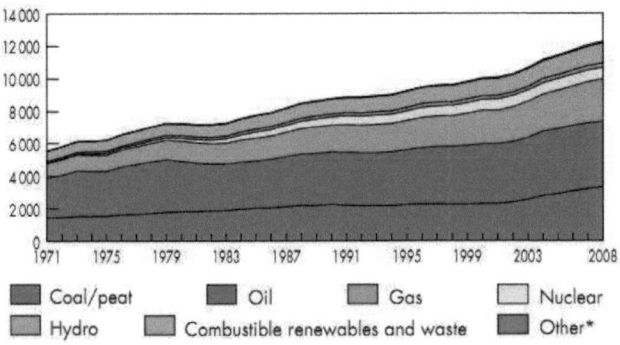

Figure 3.2: Evolution from 1971 to 2008 of world total primary energy supply by fuel (Mtoe) (IEA, 2010)

Over 70% of this growth should come from developing countries, where populations and economies are growing considerably faster than in the OECD[5] nations. China alone should account for about 30% of increased energy demand (WEC, 2007). This is one example of some of the World Energy Council Scenarios, which were published by the United Nations (UN, 2000; UN, 2004). They all assume that world population will increase from its current level of around 6 billion to 9 billion by 2050. These scenarios incorporate different assumptions about rates of economic growth and the distribution of that growth between rich and poor countries; about the choices that are made between different energy technologies and the rapidity with which they are developed; and regarding the extent to which ecological imperatives are given priority in coming decades.

Rising living standards means the consumption of more energy each year. For developing countries like China and Brazil to reach European levels in the use of energy sources would require an increase in environmental impacts of the use of fossil fuels. As Figure 3.3 shows, in 2003 Asia had overlapped the USA in regarding the CO_2 emission

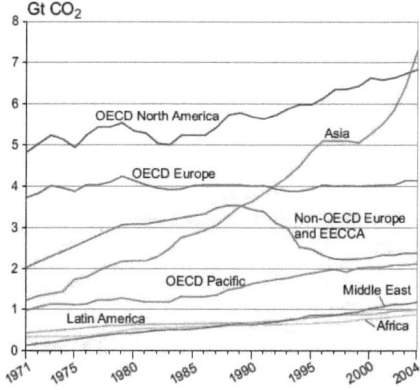

rates.

Figure 3.3: Global trends in carbon dioxide emissions from fuel combustion by region from 1971 to 2004
(Source: IPCC, 2007 a)

The global emission of gases generated from the combustion of fossil fuels is directed related to the greenhouse effect. The greenhouse effect is essential in maintaining the

---

[5] Organisation for Economic Cooperation and Development

Earth's surface at a temperature suitable for life. The greenhouse gases (GHG) are water vapour, CO_2 and methane. Since the Industrial Revolution, the burning of fossil fuels has been adding substantial quantities of CO_2 to the atmosphere in a short time, intensifying the greenhouse effect.

Biofuels should at least partially diminish GHG emissions. The IEA (2004) stated that a core concern would be the emissions released during the fuel production and use. The general outcome on net CO_2 emissions would then rely on the balance between the fossil fuel used to produce the biofuel and the fossil fuel displaced by it. However, the results depend on assumptions regarding crop rotation, fertiliser use and the use of by-products.

3.2 Political framework in Europe

Since the 1970s there has been a large expansion of biomass-based electricity generation, with an increased emphasis on generating efficiency, resulting in electricity exports into liberalised or deregulated markets. In addition, there has been an expansion of district heating schemes with CHP in Scandinavia, based on straw in Denmark and wood residues in Sweden and Finland. In countries with extensive coal-fired electricity generation there have been incentives under climate schemes to co-fire biomass in order to achieve carbon offsets of up to 15% (World Energy Council, 2007).

Every country in Europe has included bio-energy in its energetic and climatic policies. The agreed reduction in GHG emissions of at least 20% below 1990 levels by 2020, together with a 20% renewables target, was a crucial step for the EU's sustainable development and a clear signal to the rest of the world that the EU was ready to take the action required by the Kyoto protocol (EC, 2009). However, this action alone is not enough to deliver the goal of keeping global temperature increase below 2 °C compared to pre-industrial levels.

In the German electricity sector, the volume of avoided greenhouse gases totalled 74 million tonnes, around 55 million tonnes of which is electricity remunerated under the Renewable Energy Sources Act (EEG, 2007). This legislation gives priority to the feed-in of electricity from renewable energy sources into the national grid at largely fixed fees and it is crucial to the development of the renewable electricity sector.

Furthermore, around 30 million tonnes of emissions were avoided in the heat sector, and approximately 5 million tonnes in the fuel sector (BMU, 2010 a).

The growing proportion of renewable energy sources reduces emissions from the energy sector, and makes a significant contribution towards achieving the German Government's reduction targets. The use of renewables enabled Germany to avoid some 107 million tonnes of CO_2 in 2009, representing a significant contribution towards the climate protection target. Germany has undertaken to reduce its greenhouse gas emissions by 40 % by 2020, compared with 1990 levels (BMU, 2010 a). The use of bioenergy is to be further expanded. The technical potential required for this is available in Germany, where bioenergy accounted for 3.9% of total electricity consumption, 6.2% of total heat demand and 7.6% of total fuel consumption in 2007 (BMU, 2010 b).

The most important source of bioenergy in Germany is wood. In addition to forestry, agriculture also plays an important role in producing biomass for energy recovery. Already in 2007, more than 10 % of agricultural land was used to grow energy crops in Germany, with a focus on oilseed rape cultivation for the production of biodiesel and the provision of substrates for biogas production (BMU, 2010 b).

About one quarter of Germany's wood production (lower quality line of production) is used for generating energy and approximately three quarters are used as material (Kaltschmitt et al., 2009). Waste wood and used wood are also used for energy production. Therefore, the next chapters will show how biomass residues are processed and converted into energy. In chapter 2.4.7, the author will describe a case study in Germany concerning the use of biomass residues to produce heat, electricity and also biofuels (pellets).

3.3 Advantages of solid biomass

Biomass provides an alternative for changing the current scenario of energy production and consumption. It is an appealing alternative to fossil fuels because it can be found in the region where is used. Biomass can be the waste products from a community; anything ranging from food scraps, fibres or surplus waste from agriculture. All of this material can be used to create energy. This is handy not only because is cheap but also minimizes costs involved with waste disposal and it also reduces waste going to landfill (Kaltschmitt et al., 2009).

The energetic balance of biomass is positive. The amount of energy needed for the preparation of biomass fuels is lower than the energy that is gained by combustion. A variety of biomass resources are utilized through combustion, to generate bioenergy in the forms of electricity and heat. Estimations prove that the cumulative residue and organic waste could provide between 40 EJ and 170 EJ of energy per year (UCE-UU, 2001). Furthermore, the ashes obtained from biomass combustion can be used for fertilization because biomass ashes have no risk of soil contamination by heavy metals, which are often found in fossil fuels.

The potential benefits of biomass increases each year due to the increase in energy consumption patterns, followed by the shortage of the world's fossil fuel reserves. Furthermore, the transport and storage of biomass have much less environmental risk compared to fossil fuels. By using biomass, the construction of vulnerable gas and oil pipelines, which need constant monitoring, can be decreased. If the oil consumption could decrease due to the use of biomass, this scenario could help avoiding the offshore oil disasters and ocean ecosystem contamination that have occurred in the past.

The import of fossil fuels occurs often because oil and natural gas are mostly located in few regions of the globe. Middle East has 66.8% of the total proved reserves of oil and together with Russia have more than 65% of the natural gas world's reserves (BP, 2010). Furthermore, most of the reserves are in hands of few enterprises. Hence, the commercialisation of fossil fuels brings any economical benefits to the region and makes countries dependent of political instability. Since biomass fuels are mostly used in the region where is produced, its commercialisation as biofuels brings positive socio-economic outcomes.

3.4 Biomass for heat and power

During the last decades, oil and natural gas prices has been considerably increasing. Hence, European countries started investing technology in biomass power plants and household ovens. However, technology has been mostly developed for firewood. This includes heating ovens, firewood furnaces, automatic wood chip furnaces, and also decentralized biomass plants using combined heat and power cycles (CHP).

One of the major problems of current patterns of biomass use for energy is the low conversion efficiency. In European households, most biomass is burnt in so-called three stone stoves with an average conversion rate of 10%. In urban areas or larger settlements, larger biomass powered plants are common, but due to maintenance problems, low technical standards and lack of knowledge about operating them, conversion efficiencies are of the same order of magnitude of roughly 10-15%.

The efficiency of a power plant is the percentage of energy content of the fuel input that is converted into electricity output over a given time period. Efficiency of electricity generation is defined as the net quantity of electricity produced generated per quantity of fuel fired in the relevant power plant (both expressed in the same energy units). In case of boilers or cogeneration plants, the efficiency of heat generation is defined as the quantity of heat generated per quantity of fuel fired in the boiler or cogeneration plant. The average efficiency of electricity (or heat) generation refers to the efficiency over a longer time interval that is representative for different loads and operation modes, including start-ups (UNFCCC, 2010). Industrial biomass plants are estimated to operate within the same efficiency range. In industrialised countries, average conversion rates of 70-75% are common (Kaltschmitt et al., 2009).

Compacted forms of biomass such as wood pellets and briquettes can also be used for combustion. The main advantage of biomass compaction is the reduction of volume, which plays a crucial role on the logistics and on the economic feasibility of the heat and power facility. Therefore, most of the CHP plants in Europe are integrated to production/source of biomass. In some cases, the compaction facility is also coupled to the CHP plant.

The use of biomass for heat and power generation involves a complex chain of arrangements in order to provide the desired amount and shape of biomass, which should be converted into energy. Hence, the following chapter will give an overview of the types of solid biomass used for combustion, the state-of-the art of the supply chain of biomass, and the several combustion technologies.

3.4.1 Energy crops

Energy crop is the term used to define plants, which are cultivated specifically for energetic purposes. They currently contribute a relatively small proportion to the total energy produced from biomass each year, but the proportion is set to grow over the next decades. The IPCC SRES scenario estimates the realistically achievable potential for energy crops by 2025 to be between 2 and 22 EJ yr-1 (Sims et.al, 2006). Energy cropping alone is not sufficient to reduce emissions, but it is an alternative for fossil fuels and it is an important component of the energy mix for climate mitigation measures.

Agricultural crops such as corn, wheat, sugar cane, beets, potatoes and tapioca can be processed into carbohydrate feedstocks, the primary raw material for the most fermentation processes. These fermentation processes can convert those feedstocks into a wide variety of valuable products, such as bio-ethanol. Oilseeds, as soybeans, rapeseed and palm seeds can be processed into biodiesel (Soetert and Vandamme, 2009).

Forestry crops such willow and poplar in Germany (Grünewald et al., 2007) and eucalyptus in Brazil (Müller et al., 2005) have been successfully cultivated for energetic purposes. Forest thinning and removal of small-diameter, low value trees are integral parts of forest management, which aims ecological restoration and timber stand improvement. However, there is also the potential to drive unsustainable levels of harvesting, with negative consequences for biodiversity, soil, and water conservation.

3.4.2 Agricultural and forestry residues

Crop wastes and agro-industrial residues can be considered as fuel sources. The crop wastes include the biomass, which remain in the field after harvesting, for instance, straw from sugar cane, maize and wheat. On the other hand, the agro-industrial residues like, rice and coffe husks and bagasse have higher homogeneity, contain lower moisture and are concentrated on the processing plants. Practically, all these residues could directly be compacted into pellets and briquettes and converted into energy.

Forest biomass residue is the by-product of current forest management activities. It is a renewable, low-carbon feedstock that can substitute fossil fuels in the production of energy and other products (Caputo, 2009). Rests from forest decay or forest management activities, which are aimed to maintain a healthy and constant forest growth, are also used as firewood. Forestry residues that are obtained after suitable forest management, can increment future productivity of forests. However, woody residues like barks and stems are mostly rejected for marketing due to irregular characteristics for being used as a biofuel. Moreover, there is no quantification of the amount of wood residues, which are left on the forest ground after forest clearing. Hence, this type of wood is considered a "free-commodity", which would be available for collection and upgrading before it decomposes. On the other hand, the wood residues from industrial activities, such carpentry, are regular in size and moisture, becoming an important source of solid biofuel.

The conversion of biomass to generate heat and power starts with the harvest, collection, processing and transportation of the raw material. Therefore the author will describe how biomass logistics for wood sub-products and agricultural residues is functioning in Germany. The aim is to study which technologies have at most successfully been used in order to create a scenario for Brazil.

3.4.3 Biomass logistics

Biomass is transported from the field or processing locality to intermediate storage to await truck transportation. If the fields are near to the power plant (less than 30 km), the biomass can be transported directly from the field to final storage at the power plant. The storage should have the minimum level of moisture possible.

For wood harvesting the level of mechanization varies a lot. In Germany, timber felling, delimbing and forwarding can be done manually. However, harvesting short rotation crops for wood chip production today in Germany is well organized and mechanized. Hence manual labor is reduced to a minimum.

Before transportation to an on-site or off-site chipper each branch has to be bucked and moved to forest pathways in order to be collected or chipped. Mobile chippers are used on-site to produce wood chips, which are transferred to storage facilities. They are often applied in connection with motor-manual methods in private or small-scale sector, whereas stationary chippers are used for large-scale production with a higher level of mechanization in bucking and forwarding. Two methods for wood chip production are in operation: whole tree usage for small-wood and delimbing or treetop cutting for residue usage. Energy and labor effort for wood chip allocation from forests is comparable to wood log. Bottleneck for forest wood chip allocation is the labor effort for forwarding e.g. collection of the material.

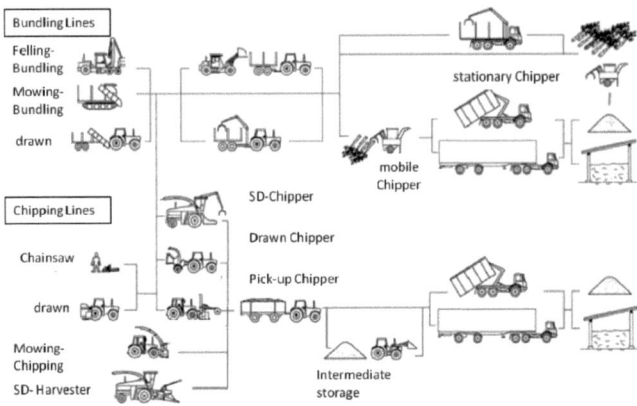

Figure 3.4: Mechanization of short rotation woody crops (adapted from Kaltschmitt et al., 2009)

Continuous harvesting and chipping is only performed with SD-Harvesters (Figure 3.4), any other method conducts interrupted harvesting and chipping. Completely mechanized wood chip production from short rotation crops is less energy consuming than from forest, due to reduced effort of bucking and machinery travel distance.

Wood chips from carpentry or industrial production residues, the so-called "white wood chip" from timber without bark, is a noble raw material for chipboard or pellet production. Therefore energetic usage seldom is applied (see Figure 3.5). The usage of "black wood chip" with bark mainly from forestry residues is common in power plants.

 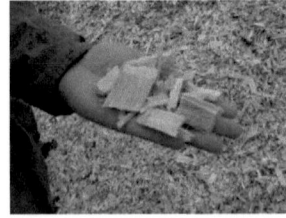

Figure 3.5: Black wood chip on the left and white wood chip on the right (own source)

For agricultural residues like straw and husks different levels of specialized machinery are available (Figure 3.6). The proper moisture for harvesting is 40% or higher and it is not suitable for combustion (Kaltschmitt et.al, 2009). Through swath drying subsequently to harvest, applicable water contents about 15% can be achieved (Kaltschmitt et al., 2009).

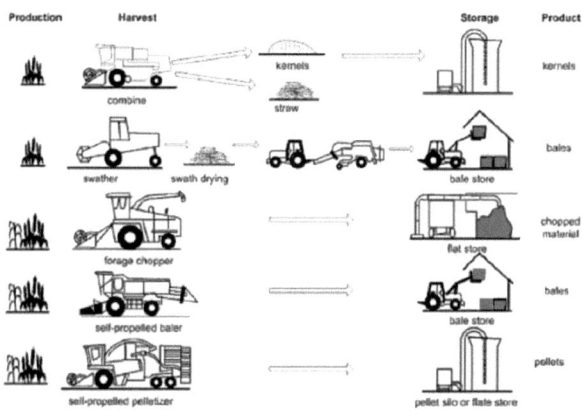

Figure 3.6: Herbaceous combustible material allocation (adapted from Van Loo and Koppejan, 2008)

In Germany, cereal and rape straw is compacted after harvest. Other crops like silage maize, green fodder cereals and Miscanthus undergo chipping after continuous harvesting (Kaltschmitt et al., 2009). However, the variety of herbaceous combustibles (agricultural residues) in Germany is rather small. The properties of the existing crops are diverse and it plays a role in determining the harvesting procedures.

Biomass can be transported in two ways: in bales or chopped. The advantage of chopped biomass is the suitability for feeding lines in power plants. Conversely the bulk density of chipped matter is lower with respect to bales, which requires costly transportation to processing facilities (Kaltschmitt et al., 2009). The preparation of bales creates material loss during harvesting. Baling has a tendency for bigger formats and mechanized handling. The bales have a cross section area up to 1.5 m² and require front loaders to handle their weight (maximum 1 ton) (Kaltschmitt et al., 2009).

Integrated harvest with pelleting or briquetting can also improve biomass' density. Transport and combustion of such fuels are easy and effective. However, biomass compaction could imply high costs when performed in non-self-sustaining facilities.

3.4.4 Biomass compaction

Agricultural and forestry residues like rice husks and wood leftovers have very low bulk densities. This results in high costs for transportation, and also complicates the processing, storage and firing. Biomass could undergo an upgrading process -for instance densification into pellets or briquettes- in order to produce solid biofuels with more uniform properties. However, pelleting and briquetting should consume the minimum amount of energy possible in order to be economically justifiable (Werther et al., 2000). The densification of materials is achieved by increasing its mass per unit of volume. Biomass densification is usually defined as compression or compaction of biomass to remove inter- and intra particle voids with expected benefits such as: more efficient storage and transportation, reduced biomass losses and moisture reduction. Moreover, it improves the physical homogeneity of the fuel and increases its total energy content (Kaliyan and Morey, 2009 and Mani et al., 2006). The densification of biomass hinders dust emissions and explosion risks.

Figure 3.7 shows different mechanisms for compaction. At low pressure, rearrangement of the particles takes place, leading to closer and denser packing, the energy is dissipated in overcoming particle friction. The magnitude of the effect depends on the coefficient of interparticle friction.

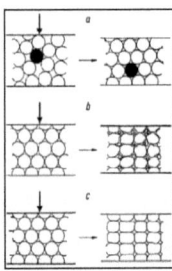

Figure 3.7: a) At low pressure, b) At high pressure with brittle material, c) At high pressure with elastic and plastic material (adapted from Pietsch 1991)

At higher pressures, the stress applied at interparticle contacts will cause fractures followed by the rearrangement of the fragments and thus a reduction in volume. At higher pressures, the elastic and plastic deformations of the particles may occur, causing the particles to flow into void spaces and increasing the area of interparticle contact (Pietsch, 1991).

Pelleting is widespread for producing animal feed and briquetting for compressing coals, minerals, and metals (Kaliyan and Morey, 2009). The process comprises pre-treatment such as grinding, drying and mixing with additives if necessary. Compression occurs through pressing the feed material through dies by rollers for pelleting or extruders for briquetting. High energy density, high strength and durability are the main aims of compression. Success mainly depends on particle size, compression force, moisture content, and on natural or added binders (Kaliyan and Morey, 2010 and Mani et al., 2006).

Figure 3.8: Straw pellets (own source)

In most industrial facilities, the compactor (pelleting mill or briquetter) is connected to a biomass drier and a biomass grinder. Some types of residues like rice husks, do not need drying, but sawdust and sugar cane bagasse, for example, have an initial moisture of more than 50%, which could hinder the densification process. Moisture can be removed from the feedstock by oven-drying or by blowing hot air over or through the particles. If the feedstock is too dry, moisture can be added by injecting steam or water into the feedstock. Large plants typically use a portion of their raw material input to provide heat for drying.

Drying and grinding biomass fuels provide significant benefits to compaction processes and boiler operation, but they must be balanced against increased capital and operating costs. The drying is considered the most costly process regarding the overall energetic balance of biomass compaction. This will be discussed in more details in Chapter 5. Most of the discussions and descriptions are based on information for woody biomass, but the information would generally apply to bagasse, husks and other types of residues.

3.4.4.1 Drying

Prior to any application as combustible or fabrication to pellets and briquettes, biomass requires drying e.g. the extraction of incorporated water. There are three requirements for drying biomass: a source of heat, a method for removing the evaporated water and some form of agitation to expose new material for drying (Lipták, 1998). In industrial scale, the moisture of biomass is diminished using driers.

The mechanisms of water incorporation are dependent on the material properties, such as fiber content and share of plant tissues. Drying is a complex equilibrium process consisting of linked mass transport, interior and across boundary surfaces, with and without phase change. The gradient and duration of drying is dependent on material properties, particle size, air properties and drying technique. For water evaporation theoretically 2.44 MJ/kg (water) have to be provided by the drying air (Kaltschmitt et al., 2009).

Freshly harvested wood can have water content up to 55%. Residues from sawmills (slab and splinter) and sawdust generally have moisture content between 25 – 40%. Residues from carpentry (sawdust and small pieces) have the lowest moisture content between 4 – 10%. For direct usage of wood log and chips, drying is recommended, preliminary to combustion but not strictly regulated (Kreimes, 2008). With respect to moisture content regulations, drying is required for pellet, briquette and wood chip production, in order to achieve maximum water content of 12 % (ZALF, 2009). The water content of herbaceous material at harvest time is between 35 – 40% and a subsequent swath drying reduces the water content to approximately 15% (Kaltschmitt et al., 2009 and Eichhorn, 1999).

Open air drying for wood requires a firm underground, regular turning and good irradiation in order to achieve water content of 20% within one day. The drying duration for split log varies between 9 and 10 month, depending on log size and bark, in order to achieve minimum water content of 15%. Freshly harvested herbaceous material has maximum water content of 40%, which often is reduced to a minimum of 15% by swath drying within a few days after harvest. For bulk goods and hay the effect of natural self-heating can be used, but no major energetic advantage has been demonstrated yet (Kaltschmitt et al., 2009).

Wet feedstocks can be dewatered prior to technical drying by using drying beds, filters and screens, presses and centrifuges. Many strategies employ waste heat for drying in order to reduce the total energy input. The share of process heat usage and extra heating is case-dependent, but is crucial for energy balancing. The common practice is to employ as much waste heat as possible for drying combustibles. However, when selecting a dryer and designing a system, it is important to consider many factors in addition to energy efficiency, such as environmental emissions, operation and maintenance costs and recovery of marketable co-products.

Biomass drying increases combustion efficiency, improves operation and can increase steam generation by up to 60% (Ross, 2008). With moist fuels, 80% excess air may be required to prevent smoke formation in wood-fired boilers. Excess air can be reduced to about 30% with dry fuel. This reduction in excess air means less heat of combustion goes into heat air. Using less excess air also reduces sensible heat losses with the flue gases, increasing boiler efficiency. Less air flow through the boiler increases the residence time in the boiler and lowers the gas velocities, aiding in more complete combustion and reducing the amount of light fuel blown out of the fire box before it completely burns.

Which drier is chosen for a particular application depends very much on the biomass properties, the opportunities for process integration (dryer and compactor) and the available environmental legislations (Amos, 1998). Drying biomass fuels has been considered as one crucial element of getting the most out of combined heat and power projects because dry biofuels increase boiler efficiency and lower atmospheric emissions. Nevertheless, economic factors may discourage the use of dried fuel.

3.4.4.2 Grinding

Size reduction is necessary before pelleting or briquetting. Common machines are hammer-, knife-, and disk-mills, and various choppers, chippers, and shredders (Yu et al., 2003). For the example of switchgrass it has been shown that the fibrous nature of grass leads to higher specific energy consumption for grinding than other herbaceous biomass (Mani et al., 2004).

The size of the fibers and particles plays a crucial role on the compaction of residues. In the case of wood, the recommended particle size is of maximum 15 mm. For other materials, it can be larger depending on which biomass is applied.

Grinding of raw biomass is carried out to reduce the size of the raw material into a desired particulate size. Grinding raw biomass can require significant amounts of energy. Shear stress and energy properties of biomass influence the size reduction process. Identifying shear response of biomass may improve knowledge to improve grinder design and reduce grinding energy (Womac et.al, 2009). On the other hand, energy consumption for biomass grinding increases with decreasing particle size.

3.4.4.3 Moistening

Moisture content or water content is described using the wet basis (w.b) and the dry basis (d.b). Those concerned with power generation most often consider moisture content on a wet basis. The wet basis moisture content directly reflects the fuel value of biomass. The moisture content of biomass on the wet basis is the weight of water in a sample divided by the total weight of the sample (McGowan, 2009):

$$\text{Moisture content (w.b) \%} = (\text{weight of water}) / (\text{total weight}) \times 100$$

Despite of the type of biomass used for compaction and of the facility, namely a pelleting mill or a briquetter, the optimum moisture content for biomass densification may range from 8% to 20% (w.b.) (Kaliyan and Morey, 2009 b) The addition of higher amounts of water can hinder the densification process, since more water cannot be absorbed by the biomass.

Water as moisture in the biomass is one of the most useful agents that are employed as a binder and lubricant. Water acts as a film type binder by strengthening and promoting bonding via van der Waals forces by increasing the contact area of the particles (Pietsch, 2005). With the help of heat, water induces a wide range of physical and chemical changes such as thermal softening of biomass, denaturation of proteins, gelatinization of starch, and solubilization and consecutive recrystallisation of sugars and salts. These physico-chemical changes affect binding properties of the biomass particles.

3.4.4.4 Briquetting and pelleting

The briquetter is a double-roll press with pockets that shape fine materials into quadratic or cylindrical bricks (Engelleitner, 1999). In most of European and German briquetting facilities, a piston extruder forms briquettes through a channel (Figure 3.9). Energy consumption of briquetting is also input dependent. The moisture content as well as material compaction and briquette shape play a major role. Additionally biomass briquetting and charcoal production sometimes are combined.

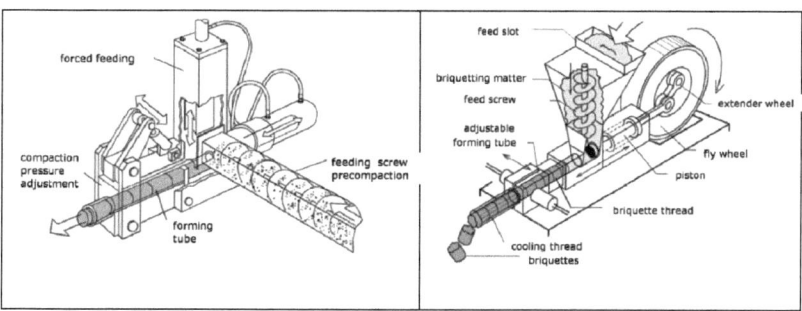

Figure 3.9: Briquetting devices (hydraulic on the right and mechanic on the left) (Kaltschmitt et al., 2009)

Pellets are formed generally in a rotating disc or drum pelletizer, which uses layering, particle growth, or agitation with the addition of water or binding agents (Engelleitner, 1999). The pelleting mill is an extrusion device that consists of a rotating ring with perforations and rollers inside it. It is used to extrude fines into cylindrical pellets. The pelleting is carried out through open holes into dies of cylindrical shape (Figure 3.10). The friction of the biomass with the metal structure causes a temperature increase inside the die openings. The heat from the pellet mill (approximately 70 °C) enables the lignin to rise to the surface of the material facilitating the agglomeration of the biomass particles.

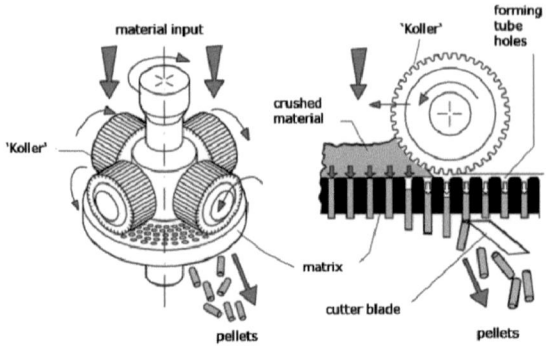

Figure 3.10: Pelleting device (Kaltschmitt, 2009)

The production of pellets can be achieved using high pressure and/or high temperature. The compaction facilities apply a pressure of 1200 Kg/cm^2 to the biomass. The influence of pressure and temperature for the compacting process depends upon the types of residue, moisture and dimensions of the particles. Uniform particle size distribution could enhance operational safety during fuel conveying into the pelleting device. Furthermore, it can help identifying whether drying or grinding of biomass is needed (Maciejewska et.al, 2006).

3.4.5 Biomass power plant technologies

Biomass can be converted for power generation using several processes. Generally, the majority of biomass-derived electricity is produced using a steam cycle process, in which biomass is burned in a boiler to generate high-pressure steam, that flows over a series of aerodynamic blades causing a turbine to rotate, which in response turns a connected electric generator to produce electricity (U.S Department of Energy, 2006).

Table 3.1: Biomass power technologies in commercial/demonstration phase

Technology Category	Biomass Conversion	Primary Energy Produced	Conversion and Recovery	Final Energy Products
Direct combustion	Stove/Furnace	Heat	Heat exchanger	Hot air, hot water
Direct combustion	Pile burners	Heat, steam	Steam turbine	Electricity
Direct combustion	Stoker grate	Heat, steam	Steam turbine	Electricity

	boilers			
Direct combustion	Suspension boilers: Air spreader stoker or cyclonic	Heat, steam	Steam turbine	Electricity
Direct combustion	Fluidized-bed	Heat, steam	Steam turbine	Electricity
Direct combustion	Co-firing	Heat, steam	Steam turbine	Electricity
Gasification (atmospheric)	Updraft, counter current fixed bed	Low Btu producer gas	Comb. boiler, steam generator, turbine	Process heat plus electricity
Gasification (atmospheric)	Downdraft, moving bed	Low Btu producer gas	Spark engine	Power, electricity
Gasification (atmospheric)	CFB dual vessel	Medium Btu producer gas	Burn gas in boiler w/steam turbine	Electricity
Gasification (atmospheric)	Co-fueling in CFB gasifiers	Low or medium Btu producer gas	Combustion turbine and steam turbine	Electricity
Slow pyrolysis	Kilns or retorts	Charcoal	Stoves and furnaces	Heat
Fast Pyrolysis	Reactors	Pyrolysis oil (bio-oil), charcoal	Combustion turbines, boilers, furnaces, reactors	Heat, electricity, synthetic liquid fuels, (BTL)
Anaerobic digestion	Digesters, landfills	Biogas (medium Btu gas)	Spark ignition combustion turbines	Heat, electricity

Source: adapted from Overend, 2003 and Broek 1995

In a furnace, biomass fuel burns in a combustion chamber, converting biomass into heat energy. As the biomass burns, hot gases are released. Mechanical power is produced by a heat engine, which transforms the thermal energy from combustion of the fuel into rotational energy. The rotational energy is transformed into electric power in a generator. A power plant includes the boiler, in case of power generation from steam, a turbine or a motor, the generator and the cooling tower (UNFCCC, 2010).

Combustion technologies convert biomass fuels into several forms of useful energy for residential, commercial or industrial uses: heat, hot air, hot water, steam and electricity (Table 3.1). Scale of biomass combustion ranges from heat supply in private households with 10 - 100 kW_{th} via district heating with plant capacities of 0.5 – 10 MW_{th} to combined heat and power plants up to a capacity of 80 MW$_{th}$ in most cases (Obernberger, 1998 and Werther et al., 2000).

Commercial and industrial facilities use furnaces for heat through a heat exchanger in the form of hot air or water. On the other hand, a biomass-fired boiler is a more adaptable direct combustion technology because a boiler transfers the heat of combustion into steam. Steam can be used for electricity, mechanical energy or heat. Steam power stations mostly operate using waste wood at low steam parameters and less complex conversion technologies (FNR, 2009).

Biomass can be fired as bales (whole or sliced), as loose material (chopped, shredded, or milled), as briquettes, and as pellets. Each of these fuel types requires specific delivery, standards and furnace technology according to the form and size of biomass (Obernberger, 1998). As wood, straw and energy crops are collected on a decentralized basis, it is feasible to use biomass in small-scale decentralized power stations.

One efficient alternative for decentralized power stations is the combined heat and power plant (CHP), which along with electricity, also decouples the resulting heat so that it can be used. Since this type of plant produces heat and electricity at the same time, the energy fixed in the biomass can be used very efficiently (FNR, 2009).

Biomass boilers supply energy at low cost for many industrial and commercial uses. A boiler's steam output contains 60 to 85% of the potential energy in biomass fuels. The conversion of solid biomass into electricity can be measured in terms of steam power process. Modern large-scale steam power stations attain steam parameters of 250 bar and 560°C, giving degrees of electrical efficiency of 43% or more.

The available alternatives for electricity production are limited by the fuel characteristics of solid biomass. Three important small scale processes today are the Organic Rankine Cycle (ORC), the steam engine and the Stirling engine. Since these have comparably low electrical efficiency, it only makes sense to use them if there is also a client for the heat produced. The ORC process and the steam engine are for instance to be found in the wood processing industry where off-cuts can be used as fuel and where both electricity and heat are required (FNR, 2009).

The ORC process makes up of organic working fluids that have better evaporation properties than water. In this case electricity production also takes place via a steam engine, albeit a somewhat modified one. ORC plants come as compact designs with an electrical capacity of 100 kWh$_{el}$ or more and an electrical efficiency of over 12% (FNR, 2009).

The steam engine can be regarded as a modern version of the classic Wattsian steam engine except for the fact that it operates in a closed cycle. As the electrical efficiency does not exceed 15%, this engine only makes sense in places where the (waste) heat is also needed (FNR, 2009).

The Stirling engine is able to convert energy into labour using generally the gas helium, which is not inflammable, via a cyclical process of changing temperature and pressure. Although the Stirling engine can run on just about any biofuel, this device has several technical constraints such as high level of particulate emissions, corrosion and heat loss. Stirling engines have an electrical output of below 10 to 40 kW$_{el}$ and some larger units can be found. Recently, the main emphasis has been laid on CHP by combining pellet ovens with Stirling engines (FNR, 2009).

There are two main types of boilers used in CHP plants and heating plants: grate boilers and fluidised bed boilers (bubbling or circulating). Both have good fuel flexibility and can be fuelled entirely by biomass or co-fired with coal (Junginger et al., 2008). However, combustion units are designed specifically for certain size ranges of fuels, for example pulverized or compacted fuels (McGowan, 2009).

Grates are usually designed to fire compacted fuels like wood chips, bark and sawdust pellets and briquettes. The grate boilers range from 10 kW$_{th}$ household heating boilers to 50 MW$_{th}$ industrial and district heating boilers. In the smallest size boiler, the grate is an inclined fire grate, but in the bigger boilers the grate usually moves mechanically. The grate (Figure 3.11) should move to prevent ash melting problems.

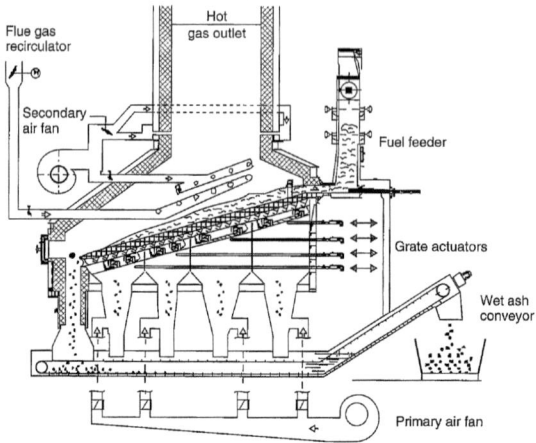

Figure 3.11: Reciprocating grate furnace (McGowan, 2009)

Capacities of grate-fired boilers range from smaller than 1 to 300 MWel in biomass-fired combined heat and power plants. Furnace temperature should not exceed 900°C during normal operation (Obernberger, 1998 and Werther et al., 2000). Advantages of grate-firing systems are insensitivity to fuel bed agglomeration, good operation at partial loads, low nitrogen oxide emissions by using advanced secondary air systems, very low dust load in flue gas, medium to low capital costs, and medium to very low operation and maintenance costs (Obernberger, 1998 and Werther et al., 2000).

The feed system of grate boilers do not suit fitting light biomass residues (low density) such as rice husks, straw or grass. An option could be mixing these materials with wood or coal and compact them into pellets or briquettes. Nevertheless, if the mixing is uneven, the light residues easily fly away with the flue gases and the fuel bed becomes cratered, leading to difficulties in controlling the combustion. Hence, for achieving a more complete combustion is recommended the use of well-distributed air supply systems and optimized grate systems (Khan et al., 2009; Obernberger, 1998; and Werther et al., 2000).

For moist fuels, approximately 80% excess air is required to prevent smoke formation, but for dry fuels, only 30% excess air is required. This reduction in excess air means less heat of combustion goes into heat air. Using less excess air also reduces sensible heat losses with the flue gases, increasing boiler efficiency. Less air flow through the boiler increases the residence time in the boiler and lower the gas velocities, aiding in more complete combustion and reducing the amount of light fuel blown out of the fire box before it completely burns.

Burning dried fuels results in higher combustion temperatures, which can be higher than the ash melting point. However, as the flame temperature increases, it approaches the fusion temperature of the ash. If the ash starts to flow and form slag, this can be very detrimental to boiler operation (Riedl et.al, 1999). Furthermore, depending on the configuration of the dryer and the boiler, and whether the dryer is a new installation or a retrofit, this may require expensive materials of construction or result in higher maintenance costs.

Fluidised bed boilers (Figure 3.12) can be considered the most modern facilities for biomass combustion but at the same time they are more expensive than grate boilers (UNEP, 2010).

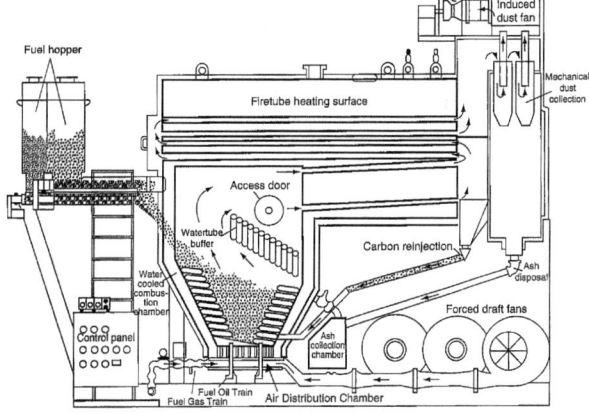

Figure 3.12: Package fluidized bed boiler (McGowan, 2009)

The costs are higher due to the special technology applied. The main advantages of fluidised bed combustion are uniform temperature distribution, large solid-gas exchange area, high heat transfer coefficients between bed and the heat exchanging surfaces, and stable combustion operation at low temperatures.

The temperature normally varies between 800 and 900°C. Nitrogen oxides can be kept low by air staging and sulphur oxide emissions by scrubbing with additives like limestone and dolomite. The technical disadvantages include agglomeration of bed materials, requirement of highly efficient gas-solid separation systems, high erosion rates of boiler internals due to high solid velocities, and high dust load in the flue gas. For dust emission reduction was recommended efficient aerosol precipitation and further treatment using electrostatic precipitators and bag house filters (Obernberger, 1998 and Werther et al., 2000).

Bubbling fluidized bed boilers can function with a variety of fuels, such as straw, wood and even sludge. In these boilers, the fuel is combusted by means of hot bed material (sand) at the bottom of the boiler. The circulating fluidised bed boilers (CFB) are also multi-fuel, functioning also with coal. In this type, bed material circulates in the boiler and fuel is fired in the bed. They are larger than the bubbling fluidised bed boilers, with a capacity of 20-550 MW (Junginger et al., 2008).

Both types of fluidized bed combustors can be used to generate hot gas for drying or for steam production. In power plants designed for dried fuel, the boiler can be smaller because less heat transfer area is needed. Second, the higher flame temperature, the more complete the combustion of the fuel. This results in more heat release, lower carbon monoxide (CO) levels and less fly ash leaving the boiler.

3.4.6 Combined heat and power generation (CHP)

Cogeneration is the combined heat and power generation (CHP). It is defined as the simultaneous production of heat and electricity from a single fuel at a sufficiently high temperature. This can enable the furnace to achieve a high overall thermal efficiency.

There are a number of cogeneration facilities fired by biomass used for power generation in the electricity sector and for heating in residential and commercial buildings. Biomass processing industries, such as the sugar cane industry in Brazil, have installed combined heat and power plants using biomass residues. Many of these have been relatively low-steam temperature installations, with only sufficient electricity to meet the plant processing needs (World Energy Council, 2007).

Small-scale CHP essentially brings the power station to the user, usually in the form of a small reciprocating engine similar to a truck engine. This will drive an electrical generator typically with a power rating between 100 kW and 1 MW output. Especially hospitals, communities and large hotels, which have a sufficiently large year-round demand for both electricity and heat, should be encouraged to invest in their own CHP plant (Boyle, 2003).

Small sized and community based plants have an increased investment costs per kW and a lower electrical efficiency when compared to coal plants. Biomass power plant efficiency is around 30% depending on the plant size (EIA, 2007). Using high-quality wood chips in CHP plants with maximum steam temperature of 540°C, electrical efficiency can reach 33 – 34%, and up to 40% if operated in electricity-only mode.

Although a small-scale CHP unit may only have an electricity generation efficiency of about 30%, less than that a conventional power station, the ability to use the waste heat makes it more energy efficient overall. The overall energy inputs and outputs of a system are shown in the Figure below.

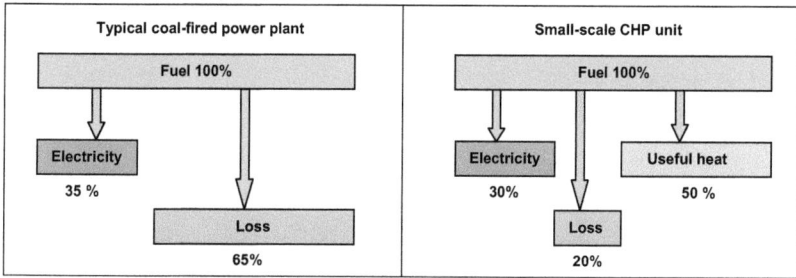

Figure 3.13: Overall efficiency of conventional CHP power plant (adapted from Boyle 2003)

The actual amount of waste heat that can be used, depends on the required temperature, but for typical heating applications the overall thermal efficiency can be over 80%. An alternative approach is to distribute waste heat as hot water from existing or adapted power stations via thermally insulated pipes to local buildings.

3.4.7 Pellet production using CHP

A sawdust pellet factory in the Lusatian region (East Germany) is able to generate 3.4 MW and produces approximately 60,000 tons of pellets per year using a cogeneration system. The heat is originated from the combustion of wood chips, which is used to dry the sawdust. This material is compacted into pellets. The availability of wood chips is influenced by the variations in the weather conditions, forestry harvesting periods, and also the market competition. The wood chips are burned with optimal moisture of 25-30%, without being previously dried. The combustion of woodchips does not cause corrosion problems in the furnaces.

The steam generated by the combustion of wood chips is transformed into heat to dry the sawdust. The sawdust is transported through bands where hot air is blown (70°C) until is dried. Sawdust is then grinded to a particle size ranging from 1-3 mm. Water is added (up to 14%) to the material in order to allow the extrusion, and depending on the tree specie, binding agents, are added. The final output are the pellets, an upgraded product, which might be sold to national and international markets.

The total electricity produced in the Lusatian pellet factory is sold to the grids. The selling price is 0.15 Euro/KWh due to the enforcement of the Renewable Energy Sources Act (EEG 2009). Through a tariff system, the German government determines a price for each unit of electricity produced from CHP plants. For new facilities with capacities in excess of 5 MW, the entitlement to grid-feed payments under the EEG Act now only applies to electricity generated in CHP plants. The bio-electricity is typically more expensive than the standard price for electricity. The government provides either the grant from its own funds or may compel transition companies to purchase the electricity produced. In this particular case, the electricity used in the pellet factory is bought from ENVIA AG for 0.07 Euro/KWh. Therefore, this tariff system enables generators like the Lusatian pellet factory to operate economically.

The production costs are approximately from 35 to 45 Euro/ton pellets and 315 kW for producing 11-12 tons of pellets per hour. The automatic packing using plastic bags accounts for almost the half of the production costs. The pellets are conditioned either in truck tanks, big bags or 15 kg bags. Approximately 85% of the pellet production is sold to wholesale dealers, the rest is commercialised to local household consumers.

The pellets produced comply with two different DIN quality standards (DIN 2008). In 2008, the typical DIN pellet was sold for approximately 125 Euro/ton and the DIN Plus for 155 Euro/ton for national and international (Denmark) wholesale dealers (Annex 1).

3.5 Discussion

As described in the Lusatian case study, electricity is bought in from all the competing power stations and then distributed through the central grid. The entrepreneurs' view is that electricity is simply a product and the aim is to maximize sales at minimum financial cost. Conventional economic theory says that the best way to do this is through the use of competitive markets and the best way to optimise such system is to leave it to the market (Green 2003).

Electricity markets differ from other commodity markets in many aspects. Demand for electricity is inelastic in the short term, storing it is impossible, components of the value chain exhibit characteristics of natural monopolies, and reliable supply has great macroeconomic importance. In the latter half of the twentieth century, these features gave rise to uncompetitive market structures in many countries. By the 1980s there was considerably pressure within the European Union to threat electricity as a totally free market commodity, especially since there was considerable cross-border trade between different European countries (Zachmann, 2009). For free marketers, the privatisation of the electricity industry was seen as a role model for subsequent similar "liberalizations" of state-controlled systems elsewhere in the world. The key ingredient was that most of the power plants became privately owned and were required to compete with each other.

Furthermore, there is the state involvement. A cheap and reliable supply of electricity is essential for economy growth. Also, new technologies such as biomass combustion may require high amounts of finance that only the state could provide. For example, in Germany there are incentives from the government to support the implementation of biomass-fired decentralized power plants. In the Lusatian case study, the entrepreneur is able to produce wood pellets using the electricity generated from wood residues. The surplus electricity he sells to the central grid, using a credit scheme, which allows the prices of his sold electricity, not be lower than the official price charged for conventional electricity.

Biomass is considered as a promising option to contribute for future energy supply. As shown along this chapter, there are different biomass power technologies in commercial and demonstration phase. Nevertheless, research is necessary in order to improve existing technologies in order to get the full potential benefits. The next chapter will show some experiments that were carried out specifically for Brazilian biomass.

3.6 References

Amos WA. (1998) Report on biomass drying technology. In: National Renewable Energy Laboratory of the U.S. Department of Energy. Available at: http://www.p2pays.org/ref/22/21209.pdf (Retrieved on July 27, 2010)

BMU - Bundesministerium für Umwelt, Naturschutz und Reaktionssicherheit (2010 a) Development of renewable energy sources in Germany. Available at: http://www.erneuerbare-energien.de/files/pdfs/allgemein/application/pdf/ee_in_deutschland_graf_tab_2009_en.pdf (Retrieved on July 2, 2010)

BMU - Bundesministerium für Umwelt, Naturschutz und Reaktionssicherheit (2010 b) General information – Biomass
Available at: http://www.erneuerbare-energien.de/inhalt/42722/ (Retrieved July 27, 2010)

BP - British Petrol (2010) Statistical Review of World Energy 2010. Available at: http://www.bp.com/liveassets/bp_internet/globalbp/globalbp_uk_english/reports_and_publications/statistical_energy_review_2008/STAGING/local_assets/2010_downloads/statistical_review_of_world_energy_full_report_2010.pdf (Retrieved on 16 May 2011)

Boyle, G., Everett, B. & Ramage, J. 2003 Energy systems and sustainability: Power for a sustainable future. Oxford University Press Inc, New York

Broek, R. van den, A. Faaij and van Wijk, J. 1995. Biomass Combustion Power Generation Technologies. Study performed within the framework of extended JOULE-IIA Programme of CECDGXII, project "Energy from Biomass: An Assessment of Two Promising Systems for Energy Production," Department of Science, Technology and Society, Ultrech University, Utrech (Report no. 95029), Available at website: http://www.chem.uu.nl/nws/www/publica/95029.htm (Retrieved on May 6, 2010)

Caputo (2009) Sustainable Forest Biomass: Promoting Renewable Energy and Forest Stewardship. Policy paper. Environmental and Energy Study Institute. Available at: http://www.eesi.org/files/eesi_sustforbio_final_070609.pdf (Retrieved on January 27, 2011)

DIN - Deutsche Institut für Normen e.V., (2008) "Terminology Database" Beuth Wissen, Germany. Available at: http://www.mybeuth.de/langanzeige/DIN-TERM/en/112289889.html&limitationtype=&searchaccesskey=main (Retrieved on July 27, 2010)

Dürrschmidt W., Zimmermann G. and Böhme D. (2006) Renewable Energies-Innovations for the future: General and Fundamental Aspects of Renewable Energies German Federal Ministry for the Environment, Nature Conservation and Nuclear Safety.

EC - European Commission (2009) Analysis of options to move beyond 20% greenhouse gas emission reductions and assessing the risk of carbon leakage. Available at: http://eur-lex.europa.eu/LexUriServ/LexUriServ.do?uri=COM:2010:0265:FIN:EN:PDF (Retrieved on Februar 1, 2010)

EEG - Erneuerbare-Energien-Gesetz (2007) Background information on the EEG Progress Report 2007. Available at: http://www.erneuerbare-energien.de/files/pdfs/allgemein/application/pdf/eeg_kosten_nutzen_hintergrund_en.pdf (Retrieved on July 27, 2010)

EEG - Erneuerbare-Energien-Gesetz (2009) http://www.eeg-aktuell.de/ (Retrieved on May 31, 2010)

Eichhorn H. (1999) Landtechnik - Landwirtschaftliches Lehrbuch. Stuttgart, Ulmer.

Engelleitner W. (1999) Update: Glossary of agglomeration terms. In: Powder and bulk engineering. CSC Publishing. AME Pittsburgh.

FAO - Food and Agriculture Organization of the United Nations (2009) State of World's Forests. Available at: ftp://ftp.fao.org/docrep/fao/011/i0350e/i0350e.pdf (Retrieved on July 9, 2010).

FNR – Fachagentur für Nachwachsende Rohstoffe (2009) Bioenergy http://www.fnr-server.de/ftp/pdf/literatur/pdf_330-bioenergy_2009.pdf (Retrieved on July 9, 2010).

Green R. (2003) Failing electricity markets. In: Utilities Policy 11 (3): 155-167

Grünewald H., Brand BKV., Schneider BU., Bens O., Kendzia G. and Hüttl RF. (2007) Agroforestry systems for the production of woody biomass for energy transformation purposes. In: Ecological Engineering 29 (4): 319-328

IEA - International Energy Agency (2004) Biofuels for transport: an international perspective. International Energy Agency. Paris, France.

IEA - International Energy Agency (2007) IEA Energy Technology Essentials. Available at: http://www.iea.org/Textbase/techno/essentials.htm (Retrieved July 20, 2010)

IEA - International Energy Agency (2009) IEA Energy Statistics. Available at: http://www.iea.org/stats/pdf_graphs/29TPESPI.pdf (Retrieved July 20, 2010)

IPCC - International Panel on Climate Change (2007 a) Climate Change 2007: Mitigation of Climate Change. Available at: http://www.ipcc.ch/publications_and_data/publications_ipcc_fourth_assessment_report_wg3_report_mitigation_of_climate_change.htm (Retrieved on July 28, 2010)

IPCC - International Panel on Climate Change (2007 b) Synthesis report. Available at: http://www.ipcc.ch/pdf/assessment-report/ar4/syr/ar4_syr.pdf (Retrieved on July 28, 2010)

Junginger M., Bolkesjo T., Bradley D., Dolzan P., Faaij A., Heinimo J., Hektor B., Leistad O., Ling E., Perry M., Piacente E., Rosillo-Calle F., Ryckmans Y., Schouwenberg P., Solberg B., Tromborg E., da Silva A., Wit M. (2008) Developments in international bioenergy trade. In: Biomass and Bioenergy 32 (8) 717-729

Kaliyan, N. and Morey V.R. (2009) Factors affecting strength and durability of densified biomass products. In: Biomass and Bioenergy 33: 337–359.

Kaliyan N. and Morey V. (2010) Densification characteristics of corn cobs. In: Fuel Processing Technology 91: 559-565

Kaltschmitt M. and Weber M. (2006) Market for solid biofuels within the EU-15. In: Biomass and Bioenergy 30: 897-907

Kaltschmitt M. (2009) Energie aus Biomasse. Springer Verlag, Berlin

Kaltschmitt M., Hartmann H. und Hofbauer H. (2009) Energie aus Biomasse. Grundlagen, Techniken und Verfahren. 2. Auflage. Springer - Verlag, Berlin Heidelberg

Khan AA, de Jong M., Jansens P.J. and Spliethoff H. (2009) Biomass combustion in fluidized bed boilers: potential problems and remedies. In: Fuel Processing Technology 90: 21-50

Kreimes H. (2008) Den Wald „verdoppeln". Heizwerterhöhung durch Hackschnitzel-Trocknung in einfachen Anlagen. Available at: http://www.carmen-ev.de/dt/portrait/sonstiges/biokraftstoffkongress08/05_Kreimes_FH_Rosenheim.pdf. (Retrieved on September 01, 2010)

Lipták, B. (1998) Optimizing dryer performance through better control. In: Chemical Engineering 105 (3) 96-104

Ludwig, J., Marufu, LT., Huber, B., Andreae, MO. and Helas G. (2003) Domestic combustion of biomass fuels in developing countries: a major source of atmospheric pollutants. In: Journal of Atmospheric Chemistry 44: 23-27, Netherlands

Maciejewska, H. Veringa, J. Sanders, S.D. Peteves (2006) Co-firing of biomass with coal: constraints and role of biomass pre-treatment. In: European Commission DG JRC Available at:
http://www.techtp.com/Cofiring/Cofiring%20biomass%20with%20Coal.pdf (Retrieved on February 01, 2011)

Mani S., Tabil L.G. and Sokhansanj S. (2004) Grinding performance and physical properties of wheat and barley straws, corn stover and switchgrass. In: Biomass and Bioenergy 27: 339-352

Mani S., Tabil L.G. and Sokhansanj S. (2006) Effects of compressive force, particle size and moisture content on mechanical properties of biomass pellets from grasses. In: Biomass and Bioenergy 30: 648–654.

McGowan T. (2009) Biomass and alternate fuel systems: an engineering and economic guide. John Wiley and Sons. New Jersey, USA

Morey V. and Doering A. (2007) Densification of agricultural biomass, bioproducts and biosystems engineering. University of Minnessota.

Müller MD., Filho AT., Vale RS. and Couto L. (2005) Produção de biomassa e conteúdo energético em sistemas agroflorestais com eucalipto, no município de Vazante, MG. In: Biomassa e Energia 2: 125-132

OHCS - The Oregon Department of Energy and Oregon Housing and Community Services (2010) Biomass Energy Technology. Available at: http://www.oregon.gov/ENERGY/RENEW/Biomass/bioenergy.shtml (Retrieved on July 27, 2010)

Obernberger I. (1998) Ashes and particulate emissions from biomass combustion - Formation, characterization, evaluation and treatment. Thermal Biomass Utilization, Vol.3, Verlag der Technische Universität Graz, Austria

Overend R. (2003) Heat, Power and Combined Heat and Power. Chapter 3 in: Sims R. Bioenergy Options for a Cleaner Environment: In Developed and Developing Countries. Elsevier, ISBN: 0-08-044351-6, 193 pages.

Pietsch W. (1991) Size Enlargement by Agglomeration. John Wiley & sons Ltd, England Otto Salle Verlag GmbH, Frankfurt am Main, Germany.

Pietsch W. (2005) Agglomeration in industry: occurrence and applications. Volume 1, Wiley- VCH Verlag GmbH, Mannheim.

Riedl R., Dahl J., Obernberger I. and Narodoslawsky M. (1999) Corrosion in fire tube boilers of biomass combustion plants. In: Proceedings of the China International Corrosion Control Conference. Paper Nr.90129 Beijing, China

Rosillo-Calle F., de Groot P., Hemstock S.L. and Woods J. (2007) The biomass assessment handbook: Bioenergy for a sustainable environment. Earthscan, London

Ross C. (2008) Biomass drying and dewatering for clean heat and power. North West CHP Application Center, Washington DC Available at: http://www.chpcenternw.org/NwChpDocs/BiomassDryingAndDewateringForCleanHeatAndPower.pdf (Retrieved on January 27, 2011)

Sims R. (2003) Climate change solutions from biomass, bioenergy and biomaterials. Available at: http://ecommons.library.cornell.edu/bitstream/1813/122/15/Invited%20Overview%20Ralph%20Sims%209Sept2003.pdf (Retrieved on January 2, 2011)

Sims R., Hastings A., Schlamadinger B., Taylor G. Smith P. (2006) Energy crops: current status and future prospects. In: Global Change Biology 12: 2054–2076. The Authors Journal compilation r. Blackwell Publishing Ltd.

Soetert W., Vandamme EJ. (2009) Biofuels. Wiley series in renewable resources. Wiley. Chichester.

UCE-UU: Utrecht Centre for Energy Research (2001) Global Restrictions on Biomass Availability for Import to the Netherlands (GRAIN). Available at http://www.uce-uu.nl/index.php?action=1&menuId=1&type=project&id=3& (Retrieved on March 3, 2009)

UN - United Nations (2000) World Population Prospects; the 2000 Revision, UN Population Division, Department of Economic and Social Affairs. Available at: http://www.un.org/spanish/esa/population/wpp2000h.pdf (Retrieved on January 27, 2011)

UN - United Nations (2004) World Population to 2300. UN Population Division, Department of Economic and Social Affairs. Available at: http://www.un.org/esa/population/publications/longrange2/WorldPop2300final.pdf (Retrieved on January 27, 2011)

UNEP - United Nations Environment Programme (2010) Technical study report on biomass fired fluidized bed combustion boiler technology for cogeneration. Available at: http://www.uneptie.org/energy/ (Retrieved on January 6, 2011)

UNFCCC - United Nations Framework Convention on Climate Change (2010) Available at: http://unfccc.int/2860.php (Retrieved on July 27, 2010)

UNFCCC - CDM - United Nations Framework Convention on Climate Change – Clean Development Mechanisms (2010) Revision to the approved consolidated baseline and monitoring methodology ACM0006 - Version 4 "Consolidated methodology for grid-connected electricity generation from biomass residues" Available at:http://cdm.unfccc.int/UserManagement/FileStorage/CDMWF_AM_I5X2LJAJB4HS YWLIIUXOKSUPQVSBGS (Retrieved on July 27, 2010)

U.S Department of Energy (2006) Distributed Energy Program, Biomass Power Fact Sheet. Available at http://www.eere.energy.gov/de/biomass_power.html (Retrieved on July 17, 2010)

Van Loo S. and Koppejan J. (2008) The Handbook of Biomass Combustion and Co-firing. London : Earthscan.

Werther J., Saenger M., Hartge E., Ogada T. and Siagi Z. (2000) Combustion of agricultural residues. In: Progress in Energy and Combustion Science 1, 27

WEC - World Energy Council (2007) Survey of Energy Resources. Available at: http://www.worldenergy.org/documents/ser2007_final_online_version_1.pdf (Retrieved on July 27, 2010)

Womac A., Ye X. , Hayes D., Igathinathane C., Klasek S., Miu P., Yang T., Yu M. (2009) Advances in Biomass Integrated Size Reduction and Separation. The University of Tennessee, First American Scientific Company, Oak Ridge National Laboratroy, University of British Columbia Available at http://www.fasc.net/advances-in-biomass.php

Yu M., Womac AR. and Pordesimo LO. (2003) Review of Biomass Size Reduction Technology. ASAE Paper No. 036077. ASABE, St. Joseph, Michigan. Available at:

http://www.biomassprocessing.org/Publications/2-Papers_presented/ASAE%20Paper%20No%20036077%20Review%20of%20biomass%20size%20reduction%20technology_yu%20et%20al.pdf (Retrieved on January 27, 2011)

Zachmann G. (2009) Empirical evidence for inefficiencies in European electricity markets – market power and barriers to cross-border trades? PhD Thesis, TU Dresden

ZALF – Leibniz Zentrum für Agrarlandschaftsforschung (2009) Standards: Entwicklung, Ziele, Anforderungen, Nutzen, Kosten. Available at: http://www.zalf.de/home_zalf/sites/innorural/downloads/Standards.pdf (Retrieved on September 01, 2010

4 Properties of Brazilian biomass residues

The use of biomass in several Brazilian regions might be an alternative for diversifying the energy matrix, which is mostly based on hydropower. The compaction of biomass could generate solid biofuels with more uniform properties. Furthermore, biomass combustion could be a cost-effective alternative for communities taking advantage of the resources available in the region to produce energy.

In Brazil, several crop residues like maize stalks and sugar cane straw serve as fertilizers and are mostly unused as biofuel. The main reasons are the difficulties in collecting them, the non-homogeneity and impurities. On the other hand, the agro-industrial residues like, rice and coffee husks and bagasse have higher homogeneity and are concentrated on the processing plants. Practically, all these residues could directly be compacted into pellets and briquettes and converted into energy. Nevertheless, the widespread mass production of compacted biomass for energy production face several limitations worldwide. The development of alternative approaches to tackle these and other limitations is still a challenge for the current research regarding the compaction and the combustion of biomass residues.

The entire pathway of utilisation of solid biofuels is characterised by the kind of biomass used, its physical features and its chemical composition. The management of the residues includes the collection, storage and transport. These steps can also involve treatment, grinding and drying of biomass before densification (Wether 2000). Agricultural residues like rice husks and wood leftovers have very low bulk densities. This results in high costs for transportation, and also complicates the processing, storage and firing. Agricultural residues could encompass a number of problems during combustion, like considerable amount of ashes. Hence, the development of technologies for using renewable resources at commercial level is needed.

The geographical, socio-economic and agricultural discrepancies from south to north Brazil have been taken into consideration during this work. The Federal State of Minas Gerais (Annex 2) is traditionally an agricultural State and most of the inhabitants of rural areas have access to electricity. Contrarily, the provision of electricity in rural north Brazil is scarce. Amazonian communities are not connected to the grid and diesel generators supply electricity. There, forestry predominates as main economic activity, followed by cattle breeding.

Based on the availability of residues in the Federal State of Minas Gerais, sugar cane bagasse, rice husks, coffee husks and sawdust from short-rotation trees were collected and transported to Germany. The aim of this chapter is to characterize agricultural and forestry residues regarding their chemical and physico-mechanical properties. The author performed an elementary analysis of different Brazilian biomass and analysed the following physico-mechanical parameters: particle size distribution and pellet stability. The investigation of the pellet stability is relevant for implementing combustion and describing the economic feasibility of the fuel.

4.1 Elementary analysis of biomass residues

The characterisation of solid biofuels can be done based on chemical and physical-mechanical characteristics. The chemical composition of biomass fuels and specially the ash content influence the choice of an appropriate combustion and process control technology (Obernberger, 1998).

The chemical characteristics includes criteria such as the concentration of certain elements (mainly C, H, O N, S, Cl, K and heavy metals); while the physical-mechanical properties primarily describe combustion related properties and mechanical characteristics able to be identified by several means including visual examination.

Main physical fuel properties are calorific value, moisture content, particle size, bulk density, ash melting behaviour (Obernberger, 1998). The calorific value first of all depends on moisture content, decreasing linearly with rising moisture content (Jenkins et al., 1998). Furthermore it is negatively correlated to the ash content. With every 1% increase in ash concentration the heating value of the fuel decreases by 0.2 MJ kg-1DM (Jenkins et al., 1996). The significance of various physical-mechanical parameters and the range of their implications are described in Table 4.1.

Table 4.1: Physical-mechanical characteristics of biofuels and their relevance (adapted from Hartmann 2004)

Parameter	Effects/Relevance
Combustion properties:	
Moisture content	Storability, calorific value, losses, self-ignition, bulk density
Calorific value	Fuel utilization, plant design
Volatile matter	Suitability for gasification, combustion behaviour
Ash content	Particle emissions, costs for use or disposal of ashes
Ash melting behaviour	Operational behaviour, furnace slagging, heat exchanges
Impurities	Ash content, ash melting behaviour
Physic-mechanical properties:	
Bulk density	Transport and storage expenditures, logistical planning
Particle density	Combustion properties (specific heat conductivity, rate of gasification)
Particle size distribution	Pourability, bridging properties, operational safety during fuel conveying, drying properties, dust formation
Bridging properties	Pourability, operational safety during fuel conveying
Stability	Dust formation, disintegration, fuel losses

The analysis of these properties for the Brazilian biomass was useful to determine whether the investigated residues could be used as biofuel. The results were compared with the European and German Standards for solid fuels. The European Committee for Standardization (CEN) and the German Institute for Norms (DIN) meet market requirements and enable efficient trading of biofuels. The DIN, together with the European Union, developed new standards for wood pellets (DIN EN 14961-2, 2010). The CEN is currently working on the improvement of standards for agricultural residues, which is not yet in force.

According to Alakangas et al. (2006), the most important technical specifications, which are in implementation, summarize the classification and specification (CEN/TS 14961) and quality assurance for solid biofuels (CEN/TS 15234). The goal is to allow producers and consumers to select the property class of biomass that is corresponding to the desired fuel quality.

4.1.1 Material and methods

Four types of Brazilian biomass were investigated: rice and coffee husks, sugar cane bagasse and sawdust from eucalyptus wood. The residues were collected in the Federal State of Minas Gerais. The coffee husks came from a large-scale farm located in the region called Zona da Mata. Rice husks and sugar cane bagasse were also collected in the same region, yet coming from small-scale agriculture. The local furniture and carpentry industry uses mostly eucalyptus wood, producing a high quantity of sawdust, which was also used on the experiments.

During the months May and June 2008, the residues were collected, stored and dried. Moisture contents of 10 to 20% after drying are the precondition for storability in order to avoid decomposition process and increase combustion efficiency (Prochnow et.al 2009). Thereafter, the samples were packed and transported to Germany with a moisture value of approximately 13%. The analysis of the transported material was carried out in August at the BTU Cottbus. This investigation covered chemical and physico-mechanical aspects, such as: moisture, ash content, heating value and so on. These preliminary results are shown in the next chapters and were important to characterize the efficiency of the pellets as a fuel.

4.1.2 Results

The water content analysis was used to determine the processing feasibility of the material. Moisture varies according to climate conditions and harvest period. Rice, sugar cane, Eucalyptus sp. wood and coffee were harvested during the dry season. After the harvest, the residues moisture varied from 14 to 30%. This moisture content could accelerate residue's decomposition process and increase its transportation costs. Hence, reducing the water content of the biomass was a necessary step before transporting the material. The biomass was exposed to air and constantly revolved for the period of approximately 6 weeks during the dry season. Afterwards, a moisture value of approximately 13% was reached and the material could be transported to Germany.

The appropriate weather conditions in Brazil associated with the storage and workforce facilities allowed biomass drying with minimum costs. However, transporting and processing biomass with high moisture could be a time, energy and money consuming activity. As Holmberg and Ahtila (2003) described, power plant efficiency decreases when firing biomass with high water content. With wet biofuel, some heat of combustion is used to evaporate the water in the fuel. Regarding dry fuel, all the heat of combustion goes into heating the air and produces the combustion.

Figure 4.1(A) shows the moisture for different types of residues. Coffee husks and eucalyptus saw wood had satisfactory moisture values after being exposed to air, as defined in the methodology. Sugar cane bagasse presented the lowest value despite its high juice content. Considering that bagasse and rice husks presented moisture varying from 7 and 9% respectively, this biomass would need a shorter drying period.

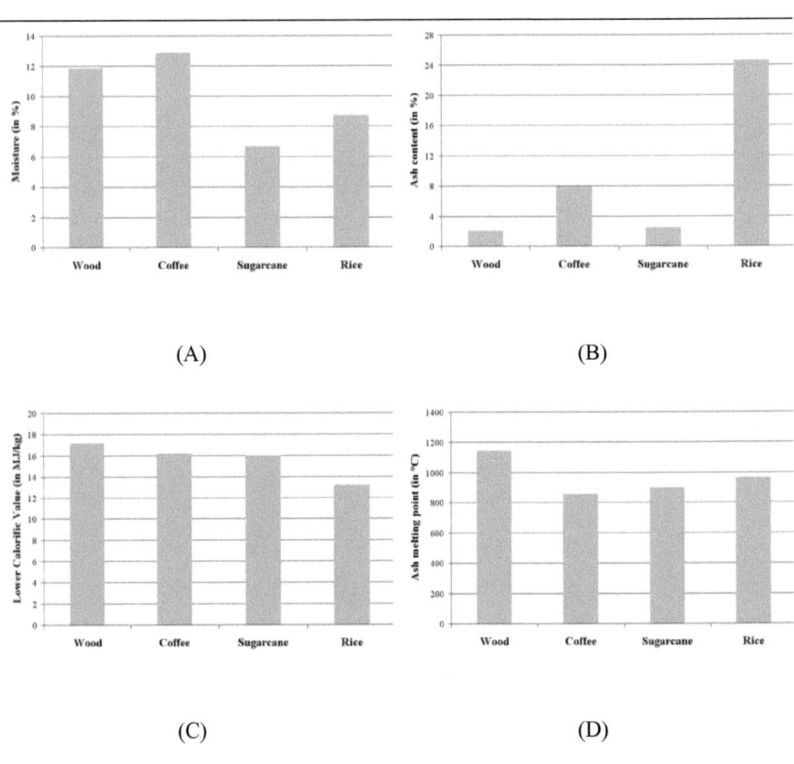

Figure 4.1: (A) Moisture content after 6 week drying; (B) Ash content (water free); (C) Lower calorific value and (D) Ash melting point (Missagia et al. 2011)

The percentage of ashes generated by the combustion process related to the total water free weight of the used material is a relevant parameter for determining pellet mixtures. The ashes contain elements like Si, Al, Fe, Cl, Ca, Mg, S, P, Na and K, which influence the ash melting behavior and deposit formation (Jenkins et al., 1998; Khan et al., 2009; Obernberger et al., 2006; Ogden, 2010). The quantity of calcium and magnesium presented in the residues can increase the ash content. Contrasting, sodium and potassium can diminish ash melting temperatures. A high ash content can decreases the heat transfer efficiency of the boiler. Furthermore, it can hinder the economic feasibility of the fuel, because removing and transporting the ashes generate high costs (Obernberger, 1998). According to the European Norm and the German Institute of Norms (DIN EN 14961-2, 2010), the ash content for solid biofuels should be up to 3 % (Annex 1). As shown in Figure 4.1(B), rice and coffee husks as single samples presented high ash contents. Nevertheless, the development of mixtures using lower quantities of these residues could be possible.

The lower calorific value shown in Figure 4.1(C) defines how much energy can be generated per amount of dried biomass. Rice husks presented the lowest heating value. This was due not only to its high ash content shown in Figure 4.1(B), but also due to its lowest quantity of elementary carbon (36.5 %) and hydrogen (6.3 %) compared to the other biomass. Contrasting, eucalyptus sawdust presented the best calorific value compared to husks and bagasse. This was due to its low ash content, and high quantity of elementary carbon (46,9%) and hydrogen (8,1%) in water free conditions compared to the other residues. According to the DIN EN 14961-2 (2010), the calorific value for compressed untreated wood should be at least 16 MJ/kg. As Figure 4.1(C) shows, the only residue that could not reach this standard was rice husks.

The ash melting point has a direct effect on the slag formation. Furthermore, the melting of ashes in too low temperatures may cause operational problems in some types of furnaces. As shown in Figure 4.1(D), saw wood presented the highest temperature. All other residues presented acceptable temperatures for ash fusion based on practical experiences with furnaces.

According to Biederman et al. (2005), wood fuels are characterised (with the exception of bark) by low ash content. On the other hand, straw, cereals, grasses and grains are characterised by considerable ash content and relatively high levels of chlorine, sulphur and alkali metals. These features were also identified in the Brazilian biomass samples.

As expected, the Brazilian sawdust had the lowest content of ash from all the tested samples. The highest concentration of ash is found in the rice husks followed by the coffee husks. Their ashes contain comparatively low Ca but high K and Si contents and therefore start to sinter and melt at significantly lower temperatures than ashes from wood fuels (Table 4.2).

Table 4.2: Ash elementary analysis of Brazilian biomass

Parameters	wet basis / dry basis	Rice husks	Sugar cane bagasse	Coffee husks	Eucalyptus sawdust
Moisture content	wt. % w.b.	8.71%	6.68%	12.86%	11.86%
Ash content	wt. % d.b.	24.61%	2.45%	7.95%	2.00%
Elements in the ashes					
Mg	wt. % d.b.	0.43	2.21	1.93	1.64
Al	wt. % d.b.	<0.05	0.20	1.09	1.88
Si	wt. % d.b.	44.60	34.90	2.47	5.06
P	wt. % d.b.	0.24	0.93	1.60	1.29
K	wt. % d.b.	2.03	5.31	36.40	27.40
Ca	wt. % d.b.	0.29	1.61	4.71	5.53
Ti	wt. % d.b.	<0.01	0.02	0.12	0.15
Mn	wt. % d.b.	0.13	0.21	0.09	0.26
Fe	wt. % d.b.	1.76	1.10	1.22	2.41
S	wt. % d.b.	0.12	0.73	1.49	1.01
Cl	wt. % d.b.	0.08	0.06	0.19	0.04

The presence of the major ash forming elements: Al, Ca, Fe, K, Mg, P, Si, Ti was also analyzed in the samples. These elements are of relevance for the ash melting behaviour, deposit formation and corrosion (Obernberger et al. 2006). The dominating ash forming elements for rice husks and sugar cane bagasse are Si followed by K. High concentrations of K and Si can result in the formation of ashes with low melting temperatures and thereby can cause slagging on the grate as well as in the combustion chamber (Dahl et al. 2004).

The guiding values for Cl and S in solid biofuels for unproblematic combustion provided by DIN (2010) were also considered for the assessment of the results of the tests carried out (Figure 4.2 and Figure 4.3).

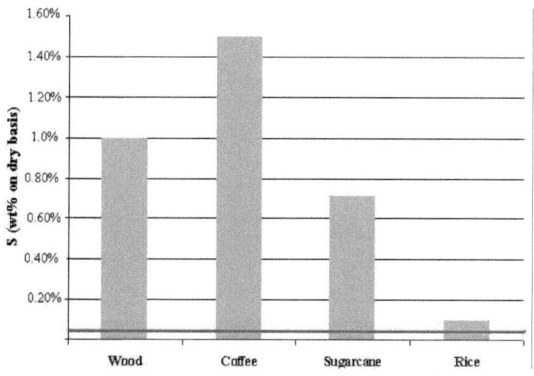

Figure 4.2: Guiding values for unproblematic thermal utilization (DIN, 2010) and sulphur concentration in biomass fuels

All of the fuel samples investigated exceeded the S guiding concentration for biomass fuels, being the coffee husks the one showing the highest levels (Figure 4.2). Sulphur is one of the elements responsible for deposit formation, corrosion, and aerosol and sulphur oxide emissions. Sulphur concentration should be lower than 0.04 % (DIN, 2010). According to the IEA (2007) there are some technological possibilities for reducing these values to the guiding ranges. These alternatives are: fuel leaching, the use of automated heat exchanger cleaning, coating of boiler tubes, appropriate material selection. On the other hand in order to reduce the SOx emissions the technological pathways suggested were dry sorption, scrubbers and fuel leaching. These measures are the same ones recommended for the reduction of HCl emissions to the guiding ranges.

Chlorine as well is involved in corrosion, deposit and aerosol formation. Furthermore it causes hydrogen chloride and dioxin and furan emissions. Chlorine content should be kept under 0.03 % (DIN, 2010). All the fuel samples analysed exceeded the threshold indicated for the levels of chlorine (Figure 4.3). In Brazil, large-scale agriculture relies on the use of pesticides and fertilizers. Pesticide formulas usually contain Cl and S, which are absorbed and stored by the plant.

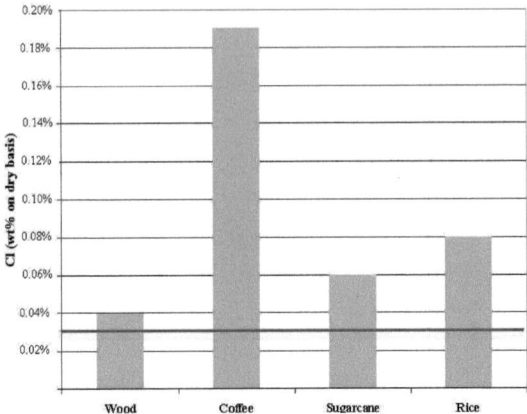

Figure 4.3: Guiding values for unproblematic thermal utilization (DIN, 2010) and chlorine concentrations in biomass fuels

Ashes from biomass combustion contain significant levels of plant nutrients, which could make the use on soils feasible. Nevertheless the ashes could be disturbed by the deposition of heavy metals on plants and soils by environmental pollution. According to Biederman et al. (2005) it is therefore recommended to separate a small and heavy metal rich side stream of the ashes (filter fly ash) from the process and to utilise the major part of the ashes produced as a secondary raw material with fertilising and liming effects on agricultural and forest soils.

4.2 Agglomeration behaviour

Agglomeration is the process of gathering fine particles into permanent larger shapes in which individual particles can still be distinguished. Hence, agglomeration is related to the particle size of the biomass and the characteristics of those particles. The characteristics of single particles are usually enhanced as the particle size decreases. When decreasing particle size through grinding, the probability of imperfections in the compacted material diminishes, resulting in a reduced risk of breakage and a higher strength in the pellet or briquette.

4.2.1 Material and methods

Due to the characteristics of the pelleting test facility and the different constitution of the materials; biomass grinding and moistening before pelleting were necessary for some types of biomass. Since saw wood is already a dusty subproduct, it does not require grinding as seen in Figure 4.4 (C). Coffee husks presented a crunchy and oily structure, Figure 4.4 (B), allowing this material be easily pressed and pelletized. Contrastingly, rice husks, Figure 4.4 (A), and sugar cane bagasse in Figure 4.4 (D) could not be pelletized in their raw form. Rice husks were often stuck in the pelleting facility due to its elastic properties. Therefore, it was necessary to grind the material. Sugar cane bagasse must be grinded due to its fibrous properties.

Figure 4.4: Raw biomass: rice husks (A); coffee husks (B); Eucalyptus saw wood (C); sugar cane bagasse (D)

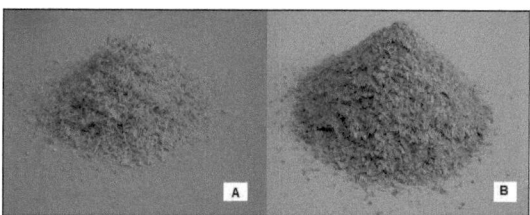

Figure 4.5: Grinded biomass (4 mm): sugar cane bagasse (A) and rice husks (B)

Grinding is a costly process, since it requires electrical energy to rotate the cutting knives. The finer the material is grinded, the more energy is consumed (Mani et al. 2004). Therefore, the maximum particle size of 4 mm was selected for the grinding process (Figure 4.5).

Raw biomass (except of sugar cane bagasse) and grinded biomass were sieved in order to obtain the particle size distribution curves. The automatic sieving device (RETSCH - AS 200/300 control) is constituted by a row of sieves with known mesh sizes placed one over another. The device is used to investigate how much percentage of each biomass sample remains at each sieve. A determined amount of material is added into the upper sieve. The machine was set to vibrate in a 10 second pulse for 5 minutes. Afterwards, the biomass in each sieve is weighted and the empty weight of the sieve is subtracted. Finally, the weight of material having a certain particle size can be calculated and compared to the whole sample weight.

Throughout the experiments, it was observed that the saw wood samples presented impurities. The material was retrieved from a small-scale carpentry in Brazil. Given that the saw wood residues have a scarce market value; the carpenters took a reduced amount of provisions during the collection and storage of this material. Hence, several screws, nails, stones and even cigarette rests were found in the samples. The cleaning of the material was carried out by the use of magnetic and air separation devices.

The densification potential of the residues was investigated by the manufacturing of pellets. The pelleting process was performed with a laboratory compactor (BEPEX - L 200 / 50 G + K) producing approximately 10 kg of pellets per hour (Figure 4.6 and Figure 4.7).

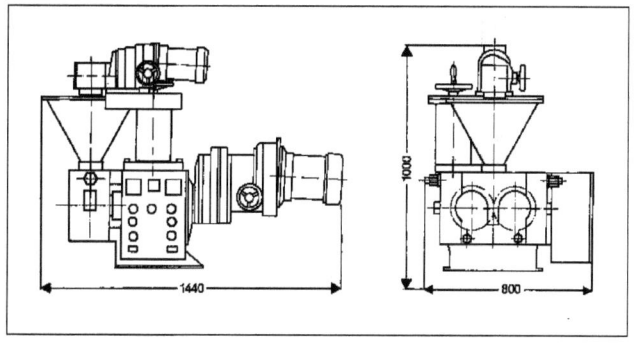

Figure 4.6: Laboratory pelleting device. Source: BEPEX - L 200 / 50 G + K

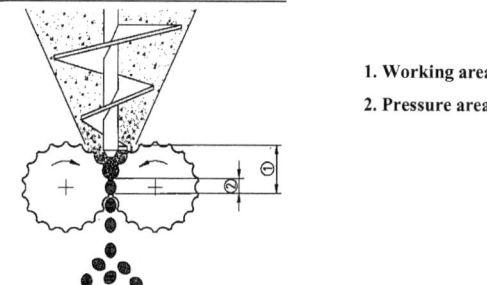

1. Working area
2. Pressure area

Figure 4.7: Extrusion of biomass residues into pellets. Source: BEPEX - L 200 / 50 G + K

Pelleting was carried out after grinding. The four types of biomass used in these experiments had sufficient natural binding due to the lignin content in their vegetal matrix. Consequently, the addition of binding agents such as starch or vegetable oil was not needed for these tests. However, moisture content was amended in between 17.5% and 18.5% in order to enable the flow of the material through the pelleting holes after the pressing. As soon as the pelleting process was completed, the pellets were spread for drying. Otherwise, there might be a risk for fungus formation. The pelleting results can be seen in the next section.

4.2.2 Results

This section presents a series of results which were obtained for analysing the viability of pelleting Brazilian biomass. The fuel properties of agricultural residues vary in a wide range. The stage of vegetation growth is the major factor to affect chemical composition as well as some physical characteristics. The raw materials sugar cane bagasse, coffe husks, rice husks and saw wood were investigated and their physico-mechanical properties are shown in the following subchapters.

4.2.2.1 Particle size distribution analysis

The particle size of biomass plays a crucial role on the agglomeration of the material. Furthermore, the particle size distribution curves could help defining which material presents more uniformity for the pelleting. As Mani et al. (2004) reported, an uniform particle size distribution could enhance efficiency during fuel conveying into the pelleting device. Furthermore, it can help identifying whether drying or grinding of biomass is needed.

As Figure 4.8 shows, coffee husks at raw form (without grinding) presented unsteady percentages of different particle sizes. Contrastingly, the Eucalyptus saw wood particle size distribution curve showed an uniform pattern, having more than 40% of its weight represented by a particle size of approximately 0,25 mm. This could be due to its previous processing in the carpentry and furniture factories in Brazil.

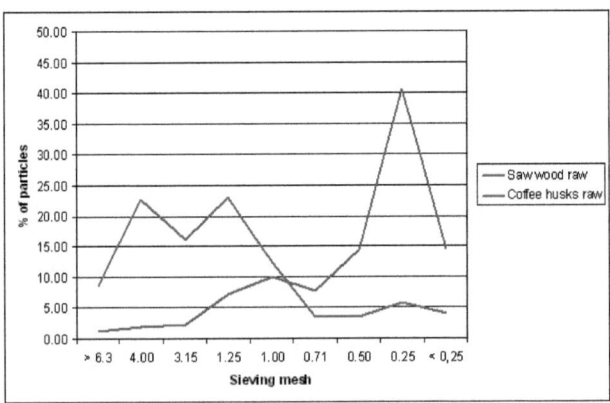

Figure 4.8: Particle size distribution curve for coffee husks and saw wood (raw form)

The grinded rice husks and sugar cane bagasse showed an uniform particle size distribution (Figure 4.9) compared to the one of the raw coffee husks. Hence, the grinding of the material could reduce the variability of the particle size distribution, facilitating the pelleting process itself and contributing to the production of pellets with more uniform density.

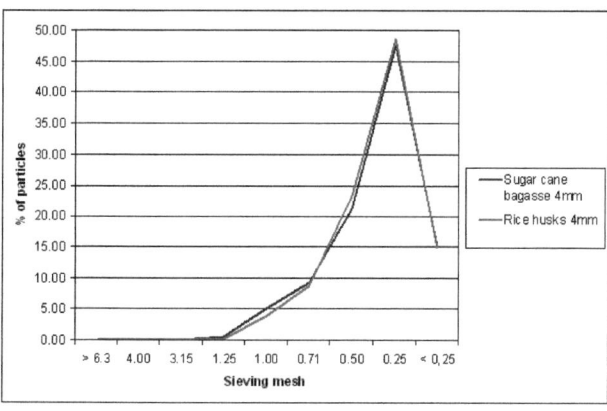

Figure 4.9: Particle size distribution for rice husks and sugar cane bagasse (4mm grinding)

Nevertheless, depending on biomass properties like moisture, oil and lignin contents, it is still possible to perform a technically feasible pelleting without previous grinding. Coffee husks and saw wood showed moisture values of approximately 12% after being exposed to air, as defined in the methodology. However, it was needed to increase its moisture content up to approximately 18% in order to be compressed in the laboratory pelleting facility. It has been tested that in large size facilities the original moisture content of the raw material is sufficient to carry out the extrusion.

The process of extrusion can be defined as the formation of cylindrical agglomerates (such as pellets) by forcing a plastic mass through a perforated die, then cutting the extruded material or allowing it to break off (Engelleitner, 1999). As Figure 4.10 shows, Eucalyptus saw wood could be pelletized on its raw form being the resulting pellets perceptibly stable.

Figure 4.10: Pellets from Eucalyptus saw wood

Considering the different dimensions of biomass upgrading, efforts should be made in order to determine the most suitable process parameters for each specific biomass assessed in order to minimize the costs of grinding while maximizing the uniformity of the final product.

The relevance of the stability assessment -for densified biofuels- lies on the fact that this parameter characterises the risk of abrasion or break ups as consequence of handling and transportation processes. It is recommended the implementation of abrasion tests in order to have a quantitative measurement of the suitability of pellets for handling and transportation. Furthermore, it is also recommended to investigate possible biomass mixtures in order to enhance not only the physical characteristics but also the chemical properties of biomass energy pellets.

4.2.2.2 Pellet stability: visually noticeable

The stability of the agglomerates under research was examined by analysing the visually noticeable stability after pelletization. Stability is a parameter that depends on the pelleting technique employed. Solid biofuels with a higher stability guarantee that a bigger part of the pellet will arrive in one piece. On the other hand, solid biofuels with less stability have a tendency to disrupt and release fine dust or particles when they are transported or stored.

The dust influences the combustion process and emissions negatively (Temmerman et al. 2004) and is considered as a health hazard while broken pellets may disturb the conveying process leading to an inhomogeneous fuel feeding during combustion.

Significant changes in the experiment results arose due to modifications in the parameters of the pelleting process such as the force applied when been pressed during the extrusion, the particle size after grinding and the moisture content.

The rice husks could be better pelletized when more force was applied during the densification process due to its elastic properties. Figure 4.11 (B) shows how even and glossy the pellet surface appears when higher forces are applied (1000 N), compared to the Figure Figure 4.11 (A) which shows pellets that had undergone the extrusion with less force (600 N).

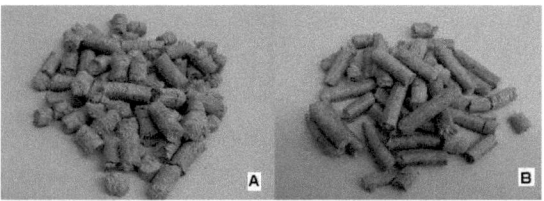

Figure 4.11: Pellets from rice husks, applying less force (A) and more force (B)

Sugar cane bagasse showed different pelleting performances when using biomass of different particle sizes. The visually noticeable stability of the 4 mm particle size sugar cane pellet was considerably less than the 2 mm particle size pellets. Figure 4.12 (B) shows how rough is the surface of the 4 mm particle size sugar cane pellet when compared to 2 mm in Figure 4.12 (A).

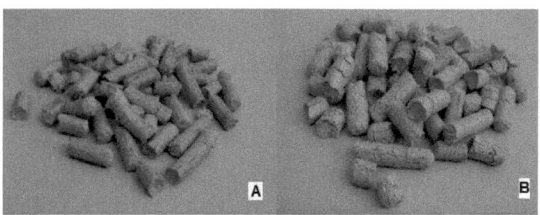

Figure 4.12: Pellets from sugar cane bagasse grinded to 2 mm (A) and 4 mm (B)

Pellets with rough surface have less stability, because they tend to release more particles and disrupt. When particles are released there is not only a fuel loss but also a risk of hindering the pellet feeding system and the combustion itself. Furthermore, a rough pellet surface has a higher propensity of absorbing water molecules from the air and pellets with high moisture values are more likely to decompose.

4.2.2.3 Pellet abrasion

Further experiments were carried out in order to analyze the effects of water content and particle size in the strength and abrasion behavior of the biomass pellets. The selected biomass was Brazilian rice husks.

The rice husks were milled with 2 mm, 4 mm and 6 mm sieve meshes. The biomass was grinded with a cutting mill (SM 2000 Retsch). The moisture content was determined using the ASAE Standard S358.2 by drying the samples in an oven for 24 hours with a temperature of 105 °C. Each experiment was repeated 5 times. The average water content of the raw rice husks assessed was 9.5%. Past experiments with rice husk pelleting showed that the ideal water content to ensure agglomeration was 17% - 20%, therefore the rice husks needed to be further moistened (Missagia et al., 2010).

Pelleting was performed with the laboratory compactor Hosokowa Bepex, Type L200/50GþK. The rice husk pellets were tested for abrasion as described in the ASABE standard S269.4. Abrasion is the loss of particulate material due to handling of pellets, like transportation and storing. The abrasion tests were carried out with the three samples of pellets (made with the raw material grinded with the 2 mm, 4 mm, and 6 mm sieves). The samples were placed in the test rotating machine (Erweka AR 401) at a rotation speed of 50 r / min for 10 minutes. The sample was then sieved with a sieve having an aperture of (0.8 * pellet diameter) mm as suggested (Narra et al 2010). The difference in the weights of the pellets before and after the abrasion test gives the abrasion value. Each experiment was repeated 5 times. On the other hand, the ability of pellets to withstand forces such as compression, impact and shear can be measured by durability experiments. Three groups of tests were carried out to investigate the hardness of pellets against compression and bending strengths. These experiments were selected as such pressures might occur during handling, transportation, and storage of pellets (Narra et al., 2010).

Compressive resistance is defined as the maximum crushing load a pellet can withstand before cracking or breaking (Kaliyan et al. 2009). The compressive stress behavior for rice husk pellets was obtained with the device (Zwick Roell-ZMART.PRO) As shown in Figure 4.13, the rectangular compression die A is moved at a constant rate from top to bottom. The pellet sample C lies on the fixed plane B. When the die A reaches the pellet C, the compression force would act on the pellet increasing the load gradually. The compression die A continues moving downward until the test specimen fails by cracking or braking. The load at fracture is read off a recorded stress–strain curve, which is reported as force.

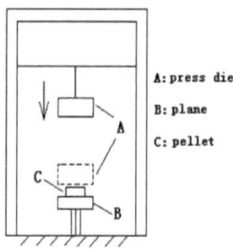

Figure 4.13: Compression stress principle

The bending strength avoid pellets' break when they experience a bending load. The "three point pressure" technique was used to measure the bending strength of the pellets. One load acts on the middle of the sample and another two forces act on the each side of sample to support it. Another bending strength test is the "point pressure" test in which the load acts on the middle of the material on the top and a plane supports the material from below (Figure 4.14).

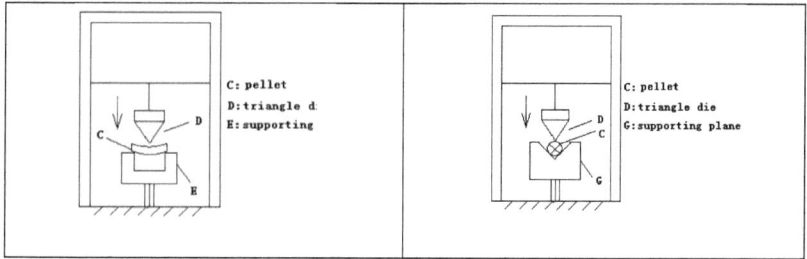

Figure 4.14: Three point pressure (left) and point pressure (right) principle

The pre-processing of the rice husks by grinding improves pellets' stability by reducing air spaces between particles during the compression, allowing closer surface-to-surface contact for a given volume of feed. Nevertheless, the results showed that the stability of the pellets was only slightly affected by the particle size. Discrepancies on the abrasion from particles sizes ranging from 2 – 6 mm grinding with 19% moisture content were observed. However, the abrasion of the pellets made of raw material grinded with the 2 mm sieve was smaller than the pellets with an average particle size of 4 mm and 6 mm with three different moisture contents. On the other hand, the water content had a considerable influence on the abrasion of rice husk pellets. The higher moisture content (for the three particle sizes) resulted in lower pellet durability.

The abrasion percentage values of the pellets with 17% of moisture content were 4% (for the pellets made with the raw material grinded with the 2 mm sieve), 5% (for the 4 mm ones) and 4.6% (for the 6 mm ones). The abrasion percentage of these three samples complies with the standard quality values of agricultural pellets as specified in DIN.

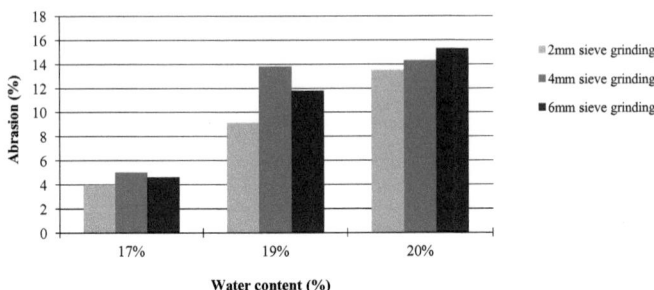

Figure 4.15: Results of the abrasion test of rice husks pellets (abrasion percentage)

Pellets with fine particle size and low water showed higher degree of hardness. Three groups of tests were made to investigate the hardness of the pellets. Figure 4.16 shows the compression stress of the pellets.

Figure 4.17 and Figure 4.18 show the bending strength of the pellets with two different methods (point pressure and three point pressure). As shown in Figures 4.16, 4.17 and 4.18 similar results were obtained when compared to the ones of the abrasion tests.

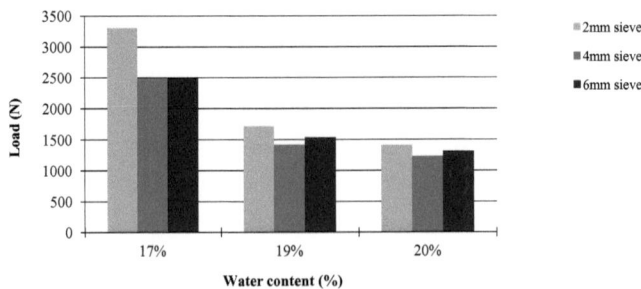

Figure 4.16: Results of the compression stress tests of rice husks pellets (load in N)

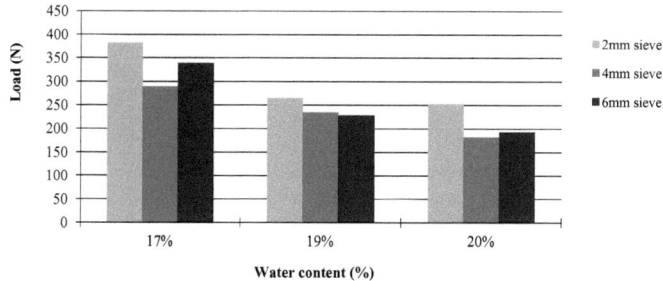

Figure 4.17: Results of the point pressure tests of rice husks pellets (load in N)

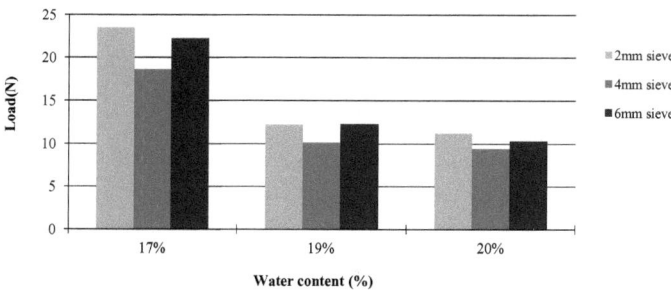

Figure 4.18: Results of the three point pressure tests of rice husks pellets (load in N)

In literature, there is disagreement regarding the impact of particle size in the durability of the final products. For instance, Kaliyan et al. (2009) suggested that generally, when biomass compaction is being carried out; the finer the grind, the higher the durability of the final product. On the other hand Grover et al. (1996) and Zahra (2006) noted that mixed particle sizes could give optimal pellet quality since they would make inter-particle bonding with nearly no void spaces. Furthermore, small particle sizes were identified as potential causes of blocking the pellet mills affecting production capacity (Shankar et al. 2010). The characteristics of the biomass, pre-conditioning processes, and densification equipment variables interact with one another (Kaliyan et al 2009). Therefore, in order to optimize the quality of densified biomass, the control of the system variables is a major requirement.

Even when the abrasion percentage of the samples analyzed comply with the standard quality values of agricultural pellets as specified in the Agro and Agro+ standards from France; they would not satisfy the norms of the Önorm M7135 (<2.3%) and DIN EN 14961-2 (2.3% - 3.5%). On the other hand it is important to note that these norms are designed for wood based pellets. Therefore, an alternative to improve the abrasion behaviour of rice husks pellets could be to mix this material with wood. Hence, more research would be needed in this regard.

4.3 Discussion

Even when all the processes aiming to produce pellets have a basic compaction mechanism in common, they can be carried out by a number of techniques. Each facility might use a particular method in order to manufacture specific types of products in respect to their diameter and length.

Some parameters were here evaluated according to the technical specifications on solid biofuels implemented by the European Union in 2010 (EN 14961-2). The standards for wood pellets became stricter than the former regulations developed by the European Committee for Standardization (CEN) and the German Institute for Norms (DIN). These standards are only defined for wood biomass. The CEN is currently working on the improvement of norms including the ones for agricultural residues pellets, which is not yet in force. The obtained pellets complied with the European norms for wood pellets, presenting a diameter of 8 mm but a variable length ranging from 40 mm to 45 mm.

According to the results, the conversion of biomass into pellets and pellets into energy could be applied for Brazilian biomass. This potential should foster research towards new power plant technologies for decentralized energy generation. However, the socio-economic feasibility of pellet production in Brazil should be taken into consideration. This will be depicted in Chapter 5.

Agricultural residues are usually widely available in Brazil. However, the ash content of these residues is relatively high compared to wood. This influences the choice of the appropriate combustion technology, deposit formation, fly ash emissions and the logistics concerning storage, utilisation and disposal (Obernberger et al. 2006). Therefore, more investigation on the combustion of blended biomass (wood and agricultural residues) should be carried out. The outcomes could result in technically and economically feasible alternatives for heat and power generation in Brazil.

It has been implied that small-scale applications for pellets or other forms of compacted biomass like briquettes allow building up a supra-regional fuel distribution system. This enables a strong competition on the fuel market. On the other hand, a disadvantage is the relatively high fuel costs and the technical modifications on the combustion devices, which should be adjusted to sharp fuel properties (Kaltschmitt et al. 2004). In this regard, it has also been documented that an expansion of the subsidies or the creation of incentives could stimulate the diversification of the alternatives of supply by stimulating the producers to develop furnaces adapted specifically to certain fuel characteristics (Coelho 2009).

The evolution of energy systems will require considerable time and expense in order to alter energy and raw material inputs, operations and products and to develop and introduce technological innovations, as well as to establish the infrastructure to support them. In order to identify possible technological-market deficits that could affect the introduction of biomass residues in the renewable energy trading schemes, the current Brazilian framework will be analyzed in the next Chapter.

4.4 References

Alakangas E., Valtanen J. and Levlin JE. (2006) CEN technical specifications for solid biofuels – Fuel specifications and classes" Biomass and Bioenergy 30:908-914

American Society of Agricultural Engineers - ASAE (1991) ASAE Standard S269.4 ; Cubes, Pellets and Crumbles - Definitions and Methods for Determining Density, Durability and Moisture Content; American Society of Agricultural Engineers: St. Joseph, MI, USA.

Biederman F. and Obernberger I. (2005) Ash-related problems during biomass combustion and possibilities for a sustainable ash utilization. Bios Bioenergiesysteme GmbH, Graz, Austria.

CEN (NORMS) - European Committee for Standardization (2011) Available at: http://www.cen.eu/cen/pages/default.aspx (Last access 8 June 2011).

Coelho S. (2009) Energy access in Brazil. Taller Latinoamericano y del Caribe: Pobreza y el Acceso a la Energía. CENBIO – Brazilian Reference Center on Biomass University of São Paulo, Santiago. Available at: http://www.eclac.org/drni/noticias/noticias/6/37496/Coelho.pdf (Last access 6 April 2011)

Dahl J. and Obernberger I. (2004) Evaluation of the combustion characteristics of four perineal energy crops (arundo donax, cynara cardunculus, miscanthus x giganteus and pacinicum virgatum). 2nd World conference and excibition on biomass for energy, industry and climate protection. Rome, Italy.

DIN – Deutsche Institut für Normen (2008) Terminology Database Beuth Wissen, Germany. Available at: http://www.mybeuth.de/langanzeige/DIN-TERM/en/112289889.html&limitationtype=&searchaccesskey=main Retrieved on: 20 April 2010.

DIN - Deutsche Institut für Normen e.V., (2010) Feste Biobrennstoffe – Brennstoffspezifikationen und -klassen – Teil 2: Holzpellets für nichtindustrielle Verwendung; Deutsche Fassung FprEN 14961-2:2010

Engelleitner W. (1999) Update: Glossary of agglomeration terms. In: Powder and bulk engineering. CSC Publishing. AME Pittsburgh.

Grover P., Mishra, S. (1996) Biomass Briquetting: Technology and Practices. Food and Agriculture Organization of the United Nations, Bangkok, Thailand. The FAO Regional Wood Energy Development Program in Asia.

Hartmann H. (2004) Physical-Mechanical Fuel Properties – Significance and Impacts, Institute for Energy and Environment, International Conference: Standarisation of solid biofuels, Germany.

Holmberg H. and Ahtila P. (2003) Comparison of drying costs in biofuel drying between multi-stage and single-stage drying. In: Biomass and Bioenergy, 26:515-530

IEA - International Energy Agency (2007) IEA Energy Technology Essentials. Available at: http://www.iea.org/Textbase/techno/essentials.htm (Retrieved July 20, 2010)

Jenkins, B.M., Bakker, R.R., Wei, J.B. (1996) On the properties of washed straw. In: Biomass and Bioenergy 10: 177–200.

Jenkins, B.M., Baxter, L.L., Miles, T.R., Miles, T.R., (1998) Combustion properties of biomass. In: Fuel Processing Technology 54: 17–46.

Kaliyan N., Morey V. (2009) Factors affecting strength and durability of densified biomass products, Biomass and Bioenergy, Volume 33, Issue 3, Pages 337-359, ISSN 0961-9534, DOI: 10.1016

Kaltschmitt M. and Hein M. (2004) Markets for solid biofuels, Institute for Energy and Environment, International Conference: Standarisation of solid biofuels, Germany.

Khan A., Jong W., Jansens P., Spliethoff H. (2009) Biomass combustion in fluidized bed boilers: Potential problems and remedies. Fuel Processing Technology. Volume 90, Issue 1, January 2009, Pages 21-50

Mani S., Tabil L.G. and Sokhansanj S. (2004) Grinding performances and physical properties of wheat and barley straws, corn stover and switch grass. In: Biomass and Bioenergy, 27 (4): 339-352.

Missagia B., Guerrero C., Narra S., Krautz H J., Ay P. (2010) Physical characterization of Brazilian agricultural and forestry residues aiming the production of energy pellets. In: 18th European Biomass Conference and Exibition, Lyon.

Missagia B., Corrêa MF., Ahmed I., Krautz HJ. and Ay P. (2011) Comparative analysis of Brazilian residual biomass for pellet production. In: Implementing Enviromental and Resource Management (Eds. Schmidt M., Onyango V. and Palekhov D.) Springer Verlag Berlin Heidelberg. ISBN 978-3-540-77567-6

Narra S., Tao Y., Glaser C., Gusovius HJ. Ay P. (2010) Increasing the Calorific Value of Rye Straw Pellets with Biogenous and Fossil Fuel Additives. In: Energy Fuels 24: 5228-5234

Obernberger I. (1998) Ashes and particulate emissions from biomass combustion: formation, characterization, evaluation and treatment. In. Thermal Biomass Utilization Volume 3, Verlag der Technische Universität Graz, Austria

Obernberger I., T. Brunner, G. Bärnthale. (2006) Chemical properties of solid biofuels—significance and impact. In: Biomass and Bioenergy: Standarisation of Solid Biofuels in Europe, Standarisation of Solid Biofuels in Europe, Elsevier Publ. Volume 30 issue 11, November, pages 973-982

Ogden C., Ileleji K., Johnson K., Wang Q. (2010) In-field direct combustion fuel property changes of switchgrass harvested from summer to fall. In: Fuel Processing Technology 91: 266–271.

Prochnow A., Heiermann M., Plöchl M., Amon T. and Hobbs P.J. (2009) Bioenergy from permanent grassland: a review. In: Bioresource Technology 100: 4945-4954

Shankar J., Wright C., Kenney K., Hess R. (2010) A Technical Review on Biomass Processing: Densification, Preprocessing, Modeling, and Optimization, Biofuels and Renewable Energy Technologies Department, Energy Systems & Technologies Division, Idaho National Laboratory, ASABE Meeting Presentation, Paper Number: 1009401

Temmerman M., Rabier F., Dugbjerg P., Hartmann H., Bohm T., Rathbauer J., Carrasco J., Fernandez M. (2004) Durability of pellets and briquettes – RTD results and status of the standardisation, Institute for Energy and Environment International Conference: Standarisation of solid biofuels, Germany.

Wether J., Saenger M., Hartge E., Ogada T. and Siagi Z. (2000) Combustion of agricultural residues. In: Progress in Energy and Combustion Science 1, 27

Zahra, J. (2006) Compaction of Switchgrass for Value Added Utilization, Auburn University, Alabama

5 Economic feasibility of biomass use for energy generation in Brazil

One of the paths towards economic sustainability refers to the availability of access to regular electricity. Such access is a key element for the economic development of communities and for the reduction of poverty (Pereira et al., 2011). It can be assumed that electricity consumption in rural areas is considerably lower than in metropolitan areas. This holds particularly for the household sector where poverty levels are higher and overall population is low (EPE, 2008).

The slow progress in expanding access to electricity is mainly due to the high costs associated to the extension of grids to remote areas and also the implementation of decentralized systems (Pereira et al., 2010). Even when there are resources available to expand the energy generation based on biomass; this needs to be done with the enhancement of socio-economic conditions of the communities involved. Therefore, this Chapter focuses on the economic feasibility of biomass energy systems, considering the social groups directly involved with its production.

The supply chain of five kinds of Brazilian biofuels (sugar cane bagasse, eucalyptus sawdust, coffee husks, rice husks and native wood sawdust) will be analysed by the author. The German experience in the compaction of biomass and conversion technologies for heat and power generation is taken as reference to identify market scenarios for Brazil.

The use of local biomass and the compaction can diminish storage and transportation costs. However, due to the variety of biomass production chains in Brazil and the type of technology applied, costs vary broadly. The author analyzes three alternatives for biomass use in different facilities: compaction into pellets and briquettes, power generation and integrated heat and power generation with biomass compaction. Several biomass supply chains were depicted in order to understand the current application of biomass residues, particularly in rural areas of Southeast and North Brazil. Both regions were chosen based on the availability of electricity for rural populations.

The several possibilities for biomass utilization have pros and contras from the economic feasibility perspective. The author proposes alternatives for several supply chains based on real parameters and real scenarios. The main objective of the economic feasibility analysis is to propose alternativess for using biomass with different outcomes, namely solid biofuels, heat and electricity. Of major interest is the compensation effect through material compaction, which allows material transportation over longer distances. The production of pellets and briquettes integrated with heat and electricity generation is considered as a promising alternative for rural communities, depending on the availability of biomass residues.

5.1 Economic feasibility: theoretical background

Financial engineering provides a variety of tools for benchmarking. The focus lies on investment appraisal, which is required to estimate the benefit of investments. An objective view of capital values can be obtained by either static or dynamic calculation strategies. Both provide information on predictable consequences of the investment in order to give an aid for decision-making. Besides technological, legal and other economical aspects this is one of the most influencing factors, apart from personal preferences.

5.1.1 Methods for investment appraisals

The static methods are easy to perform and therefore cheap. In these methods the time related to cash flow is neglected. An average of all outgoing and incoming payments is calculated to give short-term prediction or rough estimations. These methods cannot account for internal dependencies and are bound to one period (Olfert, 2008).

The *cost comparison method* is only used to compare two investments regarding their costs by calculating costs per period or per unit (Jossé, 2007).

$$\text{cost (per unit or period)} = \frac{\sum cost_{fix} + \sum cost_{var}}{total\ production\ amount\ or\ period}$$

eq. 1

The calculation considers variable and fix costs (see eq. 1) including *imputed depreciation* and *imputed interest rate*. The latter includes the costs of external finance, regardless of the origin, by a given interest rate (Jossé, 2007). The *imputed depreciation* takes the loss of value for the employed goods (material, machinery, tools, etc.) into account. Based on the real service life any good's shrinking value is calculated by means of the replacement cost at the end of lifetime (Eichhorn, 1999). Depreciation is good-dependent and can be linear (see eq. 2), declining, inclining or activity related (Olfert, 2008).

$$\text{linear depreciation (price per year)} = \frac{replacement\ cost - residual\ value}{service\ life}$$

eq. 2

The *profit comparison method* is an extension of cost comparison by incorporation of revenues. It is also used to compare two investments (Jossé, 2007).

$$\text{return (per unit or period)} = \frac{cost - revenue}{unit\ or\ period}$$

eq. 3

The *amortization method* determines the break-even-point, when an investment is paid off independently. In the static method the difference of initial investment and residual value is related to the return and depreciation per year (Jossé, 2007).

$$\text{amortization time (years)} = \frac{investment\ cost - residual\ value}{annual\ return + annual\ deprectiation}$$

eq. 4

The *leveled energy cost method* is a combination of the above listed static methods and calculates the price per unit of provided energy, based on all included investment and production costs (Quaschning, 2009). This method is appropriate in order to calculated comparable results, which serve for decision-making. Based on leveled energy costs different energy sources can be compared and identified advantageous.

leveled energy cost

$$= \frac{(\text{investment cost} + \sum_{start}^{end} \text{operation cost})\big/ \text{service life period}}{\text{provided energy per year}}$$

eq. 5

A greater significance with respect to static methods can be achieved by consideration of dates for outgoing and incoming payments. These dynamic approaches consist of several methods (present value, accumulated value, internal rate of return and annuity). However, the author considered the annuity method for the dynamic calculation of costs.

The *annuity method* calculates the annual profit factor of an investment as a product of *present value* and the capital repossession factor (*annuity factor*). The annuity factor is dependent on the service life n and the interest rate i.

$$\text{ANF}_{\text{year},i} = \frac{(1+i)^n}{(1+i)^n - 1}$$

eq. 6

$$\text{annuity} = W_0 \cdot \text{ANF}_{\text{year},i}$$

eq. 7

The *annuity method* permits evaluation of additional investments for extension and compensation or defined periods within the service lifetime. For an annuity greater or equal Zero the project can be profitable. In the latter case not less than the invested capital flows back over service lifetime.

5.2 General approach for economical feasibility of CHP plants

Generally, the design capacity of CHP power plants can be determined by heat requirements for heat-operated facilities and by electricity requirements for power-operated facilities. The main driving force is long-term stability of prices for biomass fuels. In order to achieve break-even of cost and benefit for heat-operated systems, the amount of generated heat must equal the amount of a competing system, for example heating a boiler (Steinborn, 2009).

The additional costs for a CHP facility must be compensated by electricity generation. The value of generated electricity must equal the excess investment cost for the CHP (Steinborn, 2009). The generated electricity can be applied in two ways, as feed-in to power grids or to cover own requirements. In the first case, the value of electricity is determined by the selling price. In the second case, the value of electricity is determined by the opportunity cost of electricity, for example the price which has not to be paid.

The annual load duration curve determines the amount of required heat during the year (Figure 5.1). With seasonal variations in heat demand a decrease in electricity production might occur. The example shows full load service time only for circa 30% of the whole year. The curve presents typical European energy heat demand. Nevertheless, this approach could be used for Brazilian conditions, which do not have extreme seasonal variations. In this case, the curve would present rather a constant tendency.

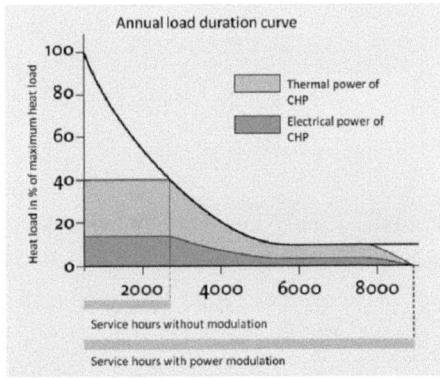

Figure 5.1: Annual load curve scheme for CHP facilities (adapted from Brauer, 2009)

According to Brauer (2009), for power-operated systems the demand for heat might not cover the supply and therefore excess heat dissipates. Full heat application can be achieved only under advantageous conditions or by design of facility clusters with included heat consumers. Generally CHP facility application requires a potential of such consumers. To run a power-operated facility cost-effective, the leveled energy costs for electricity must be covered by the market price, if no further heat application is possible. For any heat application, the revenue from heat selling has to compensate the difference between leveled electricity costs and market price. The revenue of heat in industrial clusters can be calculated by the opportunity costs of heat for the linked facility, for example the value for heat, which has not to be paid.

The energy conversion from renewable resources may be subsidized by different political mechanisms. In particular the worldwide applicable CO_2 (abatement) certificates have potential to promote positive returns, by compensating the difference between electricity production costs and return. Other subsidy mechanisms are nationally limited and therefore are not considered in the following case studies.

The emission reduction of CO_2 by the biomass power plant can be used in order to apply for CO_2 certificate trading. Specific emissions can be calculated through the IPCC method depending on fuel source (Frota et al., 2010) and after elementary analysis performed by the author (Table 5.1). The emission for the biomass fuel supply chain was determined through the GEMIS database (GEMIS, 2009) and found to be 5.0 g CO_2 / kWh.

Table 5.1: Technical and economical variables for chosen Brazilian solid biomass fuels

Fuel type (water content)	Lower calorific value	Upper calorific value	Price + transport costs	Ash content	Emissions
	MJ/kg	MJ/kg	BRL/ton	t_{ash}/ton	gCO_2/kWh [6]
Bagasse (w 12%)	16.000	18.230	100	0.245	6.6
Bagasse (w 50%)	7.860	18.230	0	0.228	6.5
Eucalyptus sawdust (w 12%)	14.830	16.650	160	0.018	6.5
Coffee husks (w 12%)	13.790	15.630	160	0.069	6.4
Rice husks (w 9%)	11.860	13.310	160	0.225	6.2
Native sawdust[7] (w)	13.920	20.930	0	0.018	6.5

---

[6] Based on IPCC method and GEMIS database

[7] Brazilian native wood mix of Amazon region (Rendeiro, 2010)

According to EPE (2010), electricity production from all generation sources in Brazil contributes to 165 g CO_2 emissions per kWh$_{el}$. Considering all sources of electricity generation as reference, the CO_2 reduction can be quantified and possible revenues from certificate trade can be determined later on. As shown in Figure 2.1, there is a large potential for CO_2 reduction using biomass power plants.

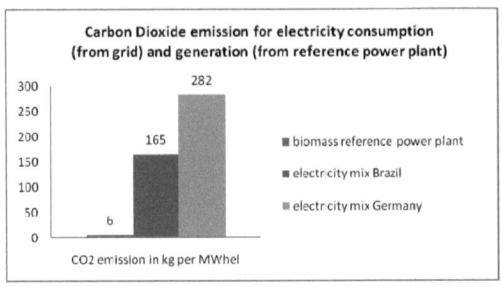

Figure 5.2: CO_2 emissions from biomass electricity compared to electricity mixtures (adapted from EPE, 2010)

Biomass fuel costs are sensitive to markets situations and are applied according to present conditions in Brazil. The calorific values are important for logistic calculations together with fuel ash content. Their amount determines required ash transportation cost. On the other hand, availability of sufficient biomass supply can be checked by means of fuel requirement calculation. The efficiency of the reference power plant determines fuel input requirements. Assuming capacity utilization above 90%, fuel supply should enable about 8,000 service hours per year (Brauer, 2009). The chosen technology for energy conversion is dependent on the prospected power plant capacity (Table 5.2).

Table 5.2: Possible conversion technologies for biomass combustion (Gaderer, 2009 and Schwarz, 2009); acreage calculated according to Kaltschmitt (2009)

Energy conversion technology	Electrical efficiency (%)	Initial investment costs (US$/kW)	Typical capacity (MW$_{el}$)	Approximate required acreage (Thousand ha)
Co-firing (CF)	35 – 40	1,100 – 1,300	10 – 50	5.4 – 27
Dedicated Steam Cycle (DST)	10 – 30 25 – 35	3,000 – 5,000	5 – 25 25 – 500	2.7 – 13.5 13.5 – 270
Integ. Gasif. Comb. Cycle	35 – 48	2,500 – 5,500	25 – 50	13.5 – 27

(IGCC)					
Organic Rankine Cycle (ORC)	10 – 20	2,500 – 3,500	0.1 – 5	0.06 – 2.7
Gas Engine (GE)	25 – 30	3,000 – 4,000	0.2 – 10	0.54 – 5,4
Stirling Engine (SE)	10 – 25	5,000 – 7,000	< 0.1	< 0.1

As shown, some technologies are only applicable for medium scale capacities, what entails large area requirements e.g. longer transportation distances and fuel prices. On the other hand technologies for small-scale applications have higher investment costs per kW and lower electrical efficiency relative to coal plants. Electricity costs in cogeneration mode (CHP) range from US$ 40 - 90 / MWh (IEA 2007).

However, according to Walter et al. (2008) Brazilian mills are producing surplus electricity from residual sugarcane bagasse since 1996 (exceeding 12 kWh per tonne of biomass crushed: the estimated electricity self consumption of a characteristic mill). Therefore, most of the Brazilian mills can operate in a self-sufficient way. According to Desplechin (2011) in Brazil, currently, bioelectricity from bagasse is funnelled back into the sugarcane mills to meet 75.5% of the sector's electric energy demands. Besides, any power surplus is sold to local electricity grids.

Due to the variety and availability of raw materials and type of technology applied, costs of biomass power plants vary broadly. The economic feasibility of biomass plants is dependent on the existing supply and process chains of raw material. Hence, it is a challenge to identify typical costs for biomass heat and electricity generation. Nevertheless it is recommended to use local biomass to avoid costly and energy-consuming transportation. Furthermore, the compaction of biomass can diminish storage and transportation costs.

5.2.1 *Biomass supply and process chain*

In the following section, the process chain of some Brazilian wooden and fibrous biomass will be described. Here the main energy consuming steps are size reduction and drying. For allocation of fibrous material as fuel after harvesting generally no pre-processing is performed. Exceptionally due to rainy weather during harvest preliminary drying might be required. For combustion whole bales or lose straw are used as input material. In Brazil, the classic biomass use in large-scale plants is sugar cane bagasse.

For wooden biomass, starting with acquisition from forests, size reduction is conducted with varying levels of mechanisation. This directly reflects in energy input and working hours for resource allocation. For harvesting wood men-hours per solid-cubic-meter vary from 3.6 mh/scm for private harvesting to 0.3 mh/scm for professional harvesting (FNR, 2007). With application of harvesting, skidding and forwarding machinery the working effort can be reduced to 0.17 mh/scm (Roßmann, 2009). This increases energy input from forest to sawmills, by factor 10 compared to manual labour (Jamieson, 2007). In Europe and Northern America due to increasing wages, manual operation is substituted by machinery. Mechanization levels in Asia and Southern America are badly recorded.

The only wooden energy source, which subsequent to harvesting requires no further processing, is wood log. For direct usage of wood chips drying is recommended, preliminary to combustion but not strictly necessary (Kreimes, 2008). Sawdust can be used directly in fluidized bed combustion without any further processing.

Material allocation costs can be derived by consideration of the process steps as listed in the following:

- harvesting
- processing (mixing, compaction, drying)
- storage and transportation

By determination of energy and labor cost for each process step a price can be calculated based on input factors or process factors. This reflects the minimum price for the biomass resource. The input factors for biomass production can vary due to a range of possible production steps and material qualities. For residues the main variations are:

- material density and water content
- amount of residues
- transportation distance
- impurities of material

Thus the supply cost for combustible materials are not fixed for one material type, but dependent on the allocation scenario. Production residues already went through a production process and therefore the expected variation of physical parameters is respectively low.

The key properties for raw material processing are water content and dimension e.g. drying and size reduction requirements. These attributes determine energetic efficiency (not applicability) of the raw material for energy source products. For example, wood from slab and splinter with a length of 2 m still is applicable for pellet production, but mechanical processing might be consuming too much energy for large-scale application. This reflects in pellet production costs, not in their energetic advantage over fossil energy sources.

In summary, production costs e.g. primary energy input for solid biomass fuels are only comparable based on raw material origin. Fuel crops consist of the full supply chain and therefore require the highest effort for allocation. The supply costs for biofuels are not fixed for one material type, but dependent on the allocation scenario. Therefore, the author picked several supply chains of biomass residues, located in specific regions, with different holders, with existing allocation and commercialization plans.

5.2.1.1 Biomass supply: coffee husks

The coffee production in Brazil can be classified into small-scale farming (l5 - 10 ha farm size) medium-scale (below 100 ha) and large-scale farming (above 100 ha). At remote small-scale farms and medium-scale farms coffee is still dried with sunlight, with subsequent peeling. But approximately 80% of these farms (INCOFEX, personal communication) transport the unprocessed coffee bean over a maximum distance of 70 km to a processing plant, at total transportation costs of 100 BRL per ton.

Figure 5.3: Dried coffee (left) and coffee husks (right). (own source)

At the processing plant beans are dried using the heat generated by the coffee husks, whereby 1 ton coffee results in 0.5 ton of husks. The surplus of husks is sold to chicken farms. The coffee exporter enterprise INCOFEX burns coffee husks to dry the coffee beans (Figure 5.4). The efficiency of the furnaces is often low, due to the equipment situation and also to the low bulkiness of the husks. The director reported that 250 kg of husks are used to dry approximately 4 tons of coffee beans.

Figure 5.4: Coffee husk furnace (on the left) for drying coffee beans (on the right)

Additionally, INCOFEX is selling coffee husks to chicken farms. These farms use the coffee residues as a substrate for chicken beds. As reported by a coffee producer in Vicosa, at the Federal State of Minas Gerais, the commercialization of husks often happens at zero costs. Several chicken farms use coffee husks for chicken beds. As an agreement, chicken farmers collect the husks at zero costs plus transportation of 100.00 BRL / ton (ca. 30 km) to the chicken farms.

After the chicken beds are ready for disposal, coffee farmers collect the enriched coffee husks with chicken excrements to be more efficiently used as a bio-fertilizer. The use of poultry litter as organic fertilizer is an attractive alternative to make production more profitable due to the total or partial substitution of the chemical fertilizers (Benedetti 2009).

It was reported by the author that large-scale coffee producers are selling the husks (BRL 60.00 / ton) to the chicken farms and neglect the enriched fertilizer. The average price of chemical fertilizers in Minas Gerais is BRL 50.00 / ton. The bio-fertilizer costs varies from BRL 70.00 – 120.00 / ton (MFRural, 2011). It was also reported that several coffee producers integrated chicken farming in their production systems.

5.2.1.2 Biomass supply: rice husks

In the region of Vicosa, most of the rice husks production supplies chicken farms. The cultivation of rice is done in small scale, mostly organically. Through the processing of rice 30% residues arise. Several chicken farms use rice husks for chicken beds. As an agreement, chicken farmers collect the husks at zero costs plus transportation of 100.00 BRL / ton (ca. 30 km) to the chicken farms. Simirlarlly to the coffee husk supply chain, farmers collect the enriched rice husks with chicken excrements to be used as fertilizer.

5.2.1.3 Biomass supply: eucalyptus sawdust

The author investigated 6 small-scale sawmills in Southeast Brazil: 5 in the Federal State of Minas Gerais and 1 in the Federal State of Espírito Santo, and found out that most of the sawdust is sold to chicken farms, to be used in chicken beds. However, according to Walter et al. (2009), only in São Paulo, there are approximately 5 thousand pizza-houses and 8 thousand bakeries, in which ones 70% of energy source is charcoal or firewood, which is being gradually changed to wood briquettes.

Figure 5.5: Sawmill in Espírito Santo (own source)

5.2.1.4 Biomass supply: Amazonas sawdust

In North Brazil, the author investigated 3 communities in the Federal State of Pará (Amazon), which produce sawdust and use for the generation of heat and electricity. The investigated region is rich in water resources. Hence, most of the population uses canoes and motorboats as main transport. The transportation of wood logs for the sawmills is done using the river systems (Figure 5.6).

Figure 5.6: Wood log transportation in Amazonas

The logs are kept in water before milling. Once wood is removed from water, it needs to be used subsequently in order to avoid biomass decomposition, caused by high temperature and moisture of the air.

5.2.1.5 Biomass supply: sugar cane bagasse

The Organic Sugar Cooperative visited by the author cultivates sugar cane and produce organic sugar using bagasse for generating heat, which is used on the processing of sugar cane juice. In total, the association produces approximately 700 tonnes of sugar cane per year. From this amount, 200 tonnes of bagasse is obtained and used for cooking the sugar cane juice and producing brown sugar. The organic sugar is freighted with trucks to Belo Horizonte, the capital of Minas Gerais, and sold in rural markets. The transport cost 100 BRL / ton of sugar.

Figure 5.7: Sugar cane harvest (on the left) and milling (own source)

Figure 5.8: Sugar cane bagasse (on the left) and bagasse combustion for cooking sugar cane juice (own source)

Table 5.3 shows the factors that influence the allocation of biomass in the initial phase of utilization and also the existence of a market chain. The data regarding the production of residues per year, transportation distance to processing plants was collected on the facilities visited by the author.

Table 5.3: Investigated biomass supply chains in Brazil and biomass properties

Residues	Origin	Ton / year	Transport	Impuritie	Market	Destination
Coffee husks	Minas Gerais	300	70 km	No	Yes	Heat generation Chicken farms
Rice husks	Minas Gerais	50	30 km	No	Yes	Chicken farms
Eucalypt sawdust	Minas Gerais	1000	20 km	Yes	Yes	Chicken farms
Amazon sawdust	Pará	4,500	Fluvial	Yes	Yes	CHP Briquet facility
Bagasse	Minas Gerais	200	0 km	Yes	No	Heat generation

* Coffee husks (Saenger et al., 2001); rice husks (Mansaray and Ghaly, 1997), eucalyptus sawdust (Higashikawa et al., 2010); Amazonas sawdust (Rendeiro, 2010) and bagasse (Wade, 2004).

Generally, the biomass residues present low bulk density. Hence, farmers and holders of the biomass business should care about the conditioning of residues, in order to avoid moisture increase and changes in the physico-chemical properties. Moreover, as greater the volume, as costly is the allocation of this material.

The compaction into pellets or briquettes could enhance bulk density and improve the physical characteristics of the fuel. The coffee husk fuel can save storage and decrease risks of biomass ignition. Hence, it could be used direct in the furnaces for drying the coffee beans. Another option could be the commercialization of the coffee husk pellets to the chicken farms. The transportation of a compacted material decrease the allocation costs and inserts a new product on the agricultural market. However, compaction facilities demand investment costs, which in most of the cases cannot be decked by smallholders.

5.3 Biomass utilization alternatives for Brazil

The main cost factors influencing the implementation of biomass projects are listed in the following:

- Investment costs of biomass plants
- Investment costs for power grid and heat grid connection/construction
- Investment costs for working media allocation (water, electricity, diesel, etc.)
- Running costs for operation (fuel, fresh water, fuel, chemicals, disposal cost, waste water, electricity)
- Running costs for personal (salary, social service, tools)
- Other costs (maintenance, insurance, fees, rent, office)

The initial investment costs are referred as capital costs. They are determined by the credit amount, which results from planning and construction cost and the common interest rate. By means of the annuity method, within a given service life period, the annual payment to the credit grantors (annuity) can be calculated. By determination of running cost factors and all other costs, the annual payments can be opposed to the annual return from electricity and heat selling (cost comparison method). The analysis of running and other costs is performed statically. Only capital costs are considered dynamically assuming a constant interest rate. This approach calculates the present facility revenue. The results can be used for further economical calculations with dynamic methods, which are not applied because of lacking information on future interest rates and price changes.

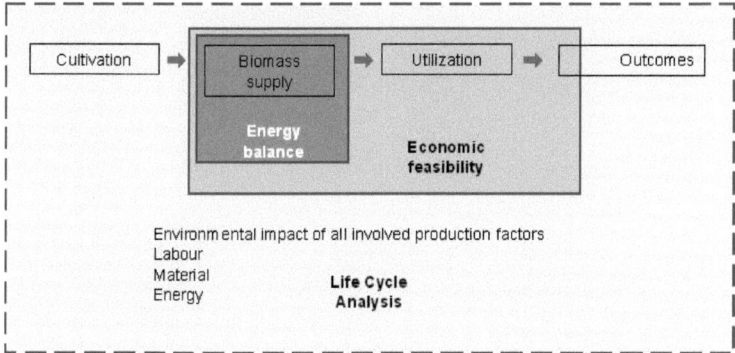

Figure 5.9: Analysis approach for agriculture and forestry residues and production residues (own source)

The economic feasibility of energy generation from biomass consists of two general considerations. Initially, the costs for energy conversion in the utilization phase have to be calculated (Figure 5.9). This involves facility planning and construction, running costs, personal costs and other facility related cost factors.

The most influencing financial factors are initial investment costs, fuel cost and market price. The investment costs are dependent on availability and advance of the applied energy conversion technology. The market price for electricity and heat may vary regionally and temporarily in a manageable range. The fuel costs are dependent on nominal market prices e.g. demand for the biomass resource or supply costs. Therefore these three factors can determine cost-effectiveness alone and have to be tuned in order to achieve positive returns.

Generally, the price for fuels is included in running costs. But for agriculture, forestry and production residues in Brazil there are often no market price. By means of allocation cost analysis an input price can be determined. This cost reference can be transformed into a nominal fuel price by predictive market estimations.

The nominal fuel price can be used in order to calculate cost functions for several scenarios for biomass fuel application. Three main approaches can be applied: the distribution of compacted biofuels from production centers (Figure 5.10, compaction case); the concentration of biomass at supply hotspots exclusively for electricity generation (Figure 5.10, power generation case) and biomass for heat and electricty generation with the production of compacted biofuels (Figure 5.10, integrated case).

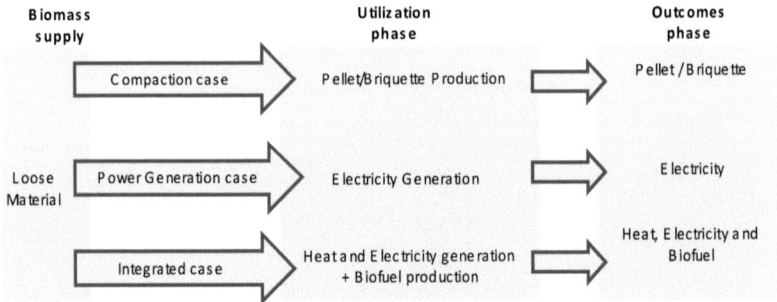

Figure 5.10: Selected scenarios for biomass utilization in Brazil (own source)

5.3.1 Compaction case for biomass residues utilization

The compaction case presents a scenario for the valorization of biomass through the production of compacted biofuels (pellets and briquettes), independent from integrated energy generation in CHP plants. The end of production chain is the biofuel at the production facility site. The factors influencing this cost calculation are: transportation, processing (grinding and drying) of biomass and electricity input for the pellet and briquette production. The compaction case distinguishes three arrangements in pellet facilities:

- Regional pelleting/briquetting facility
- Local pelleting/briquetting machines
- Mobile pelletizer

Biomass compaction refers whether to regional or to local production of biomass pellets and briquettes. Primarily the regional approach involves a production facility with industrial capacity. The design of the facility depends on the available amount of residues. A maximum transport distance of 30 km has been set in order to keep transportation costs low. The discharge of regional pelleting facilities exceeds 1 ton per hour.

The local approach refers to transportable machines with a lower capacity and a small discharge rate, with maximum 300 kg per hour. These machines require an appropriate input material condition with suitable particle size and water content. The operator is responsible to fulfill quality requirements. Production of pellets generally requires understanding of cohesion mechanisms or at least defined routines. The quality of produced pellets, in particular abrasion resistance, may vary due to input material quality variations and difficulties during the production process.

The production of briquettes is easier to handle for semi-skilled workers. Defined routines for input particle size and water content are necessary, but the process is less sensitive to input material quality variations, compared to pelleting. Even manual briquetting is possible, with a hand-operated device. The reported discharge with 300 kg per day is far below mechanical devices and only suitable in low labor cost regions, but at much smaller investment costs of € 1,500 (Gladstone, 2009).

Another option to save investment cost is the application of mobile pelletizing machinery. This device is operated with a diesel generator and can be considered a cutting edge technology. The maximum capacity is suitable to large-scale and the device is mobile, additionally it also applies for small-scale production. As rule of thumb for economic feasibility in Germany, an annual production of 1,000 tons has been reported to be sufficient (Kraft 2011, personal communication).

Additionally, all machinery types distinguish in quality, durability and maintenance service. The maintenance costs are subject to skill level of operators, raw material quality and type, climatic and installation conditions. Generally, the durability of briquetting devices has been reported to be more long-lasting than pellet devices (Wornath 2011, personal conversation). Spare parts, such as flat matrices and "Koller" for pelleting devices have been indicated as cost intensive equipments (Seipp 2011, personal communication).

5.3.1.1 Methodology for compaction case calculations

Through field research and personal conversation, the author performed a selection of compaction machinery for pellets and briquettes. All nominal data, given by technical description and list prices have been used, in order to compare costs (see cost comparison method) of the single devices. The difference of shipping cost is considered to be marginal and is not discussed further.

Table 5.4: Input data for cost comparison method

Biomass compaction	Unities
Minimum production capacity	kg/hour
Maximum production capacity	kg/hour
Required electrical power	kW
Total investment costs (nominal)	BRL

Based on investment costs, energy consumption costs and wage costs, comparison of the devices had been calculated, with a nominal output of 1 ton compacted material. For financial considerations the input material price has been left out, assuming an equal price for all production facilities. Anyway raw material costs do not refer to the compaction process itself and therefore could be externalized and considered later. In Table 5.5, several compaction machines have been listed.

The energy costs were calculated based on machine power, service hours and electricity price. Here the maximum electrical power requirement was taken from technical data sheets (see Table 5.5), weighted with service hours and an average electricity price of 350 BRL/MWh (ANEEL, 2011).

In case of fully equipped devices (industrial pelleting facility and mobile pelletizer) crushing/drying cost are included. For other machinery they are added as external electricity cost. The cost calculation for pelleting and briquetting devices involves crushing/drying cost per ton, based on the industrial scale with no heat integration. Due to strong dependency on raw material input quality and/or particle size distribution, here maximum costs per ton had been assumed for all machinery, in order to achieve a basis for comparison. Finally the maximum production output during, with maximum annual service hours results in a nominal maximum output. This is weighted with the related maximum energy cost, in order to calculate specific energy cost.

Table 5.5: Description of compaction machineries for further cost comparison analysis

Pelleting machinery	Installed Power	Company	Required workers	Man-hours/ton
Industrial pelleting	260 kW	AGICO	3	3
Mobile pelletizer (for hire)	220 kW	BauerPower	1	1*
Local pelletizer	30 kW	MABA	1	3,3
Local pelletizer	7,5 kW	PP	0,25	1,7
Local pelletizer	4 kW	PP	0,25	2,5
Local pelletizer	3 kW	PP	0,25	4,2
Briquett machinery	**Installed Power**	**Company**	**Required workers**	**Man-hours/ton**
Industrial briquetting	22 kW	WEIMA	1	2,9
Local briquetting	7,5 kW	GROSS	0,25	1,6
Local briquetting	4 kW	WEIMA	0,25	6,25
Local briquetting	3 kW	RUF	0,25	6,25

By assuming 8,000 annual service hours within a 10 years lasting lifetime of each device, the linear description is calculated and included to annual capital cost. This approach implies no external debt, for example a completely bar payment of the device at production start-up.

Generally, two cases of production are distinguished. Primarily, the regional approach requires raw material transportation to the industrial pelleting facility. Another option is transporting the mobile pelletizer to the raw material. For both transportation options distances are assumed 30 km and cost are calculated with a referred transport price. The mobile pelletizer can travel a maximum distance of 60 km, but this is strongly dependent on the conditions of Brazilian roads. Secondly, the local approach uses biomass from the vicinity and requires no transportation.

The author considered that for all machineries, the operation of the devices is performed by workers, who are granted with a minimum salary of 500 BRL/month in Brazil. Only in the case of the mobile pelletizer (for hire) the German wage is applied for comparison purposes. It is important to notice that the mobile pelletizer is until now not employed in Brazil. Hence, the German conditions were applied for specific costs calculations. Later on, another approach for a mobile pelletizer scenario for Brazilian conditions (community approach) is described aiming potential future application. The technical description of the considered devices includes worker effort considerations and serves as basis for calculation of wage cost.

The worker wage cost for the mobile pelletizer (for hire) is considered to be the 24-fold of Brazilian minimum salary within the German entrepreneurial case. According to an existing business example, production costs split into nominal wage cost / hour and production cost / ton. The calculated production cost of 70 EUROS / ton cover the expenses on energy and wage in the case of regular production. In case of irregular production, for instance due to the low quality of biomass, production capacity decreases and therefore machine operation time increases, when producing 1 ton of pellets. Therefore, splitting the costs into 35 EUROS / hour for wage and 35 EUROS / ton for energy has proven to be applicable. The application of this approach (machine for hire) has been reported to trigger a learning effect, which promotes proper preparation of raw material (Seipp 2011, personal communication).

The entrepreneurial mode of operation includes business objectives, with a benefit to the company owner. It is completely handed over to the customer and is represented as capital cost when hiring the device. The mobile pelletizer could be applied in Brazilian communities by assuming machine usage at cost coverage. The case is balanced by applying capital and wage cost approaches like for the other related machineries (Table 5.5). Capital costs calculate as linear description from investment costs for the service lifetime. The wage costs are based on working effort and Brazilian minimum salary. The energy costs are calculated by means of required energy input (technical description) and Brazilian fuel (diesel) prices.

The calculation of overall production costs per unit (see cost comparison method), whereas 1 ton of pellet or briquette is produced, allows a first analysis of different organization strategies. The considerations are made to evaluate the production cost for pellets or briquettes, when leaving the production facility and enable modular application of the results.

5.3.1.2 Results

The specific production costs for related compaction machineries were calculated and the results differentiate considerably. The analysis was performed for pellet facilities and briquette facilities separated. The capacity of biofuel production for the investigated machinery is fixed and related to costs. For higher capacities different price changes have to be considered. Normally, investment costs for local machinery double with 2-fold capacity. Here, investment costs are considered capital costs.

For industrial machinery investment costs can be considered as almost linear with capacity. Although parallel usage of preparation or feeding devices is possible, the core parts, such as 'Kollergangpresse' (for pelleting) and compactor (for briquetting) need to be doubled for 2-fold production capacity. As these parts contribute to the main investment and main energy costs, the resulting capital and energy costs marginally deviate from linear increase with capacity (WEIMA 2011, personal communication).

Figure 5.11shows the costs of pellet production from a free of charge biomass residue. This was calculated based on the production of 1 ton of pellets. For regional applications the transportation costs to the facility or to raw material (mobile pelletizer) account to a large share of the overall cost. The transportation costs for the mobile pelletizer in a community operation mode is slightly higher than for entrepreneurial operation. For local machinery preliminary crushing and grinding are considered as external costs with a maximum value. Cost basis are the pre-treatment energy and wage costs of an industrial process. For local application prices increase with decreasing capacity due to constant wage cost and lower output.

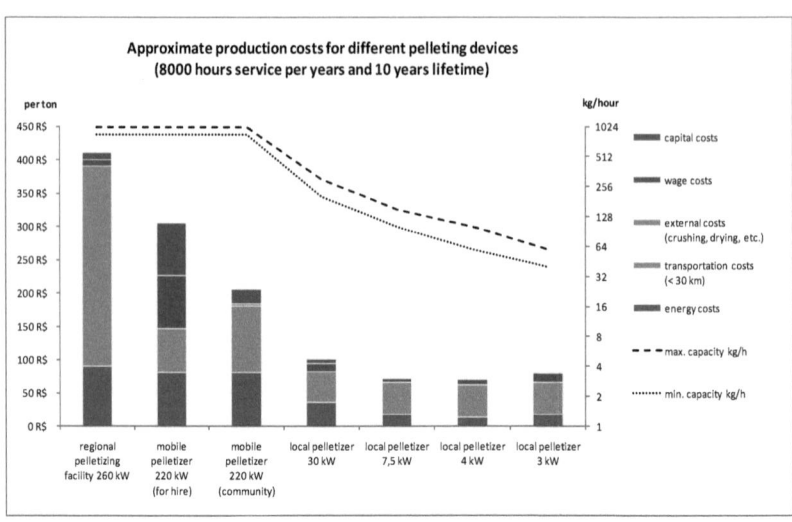

Figure 5.11: Comparison of production costs for different pelleting approaches.

Figure 5.12 shows the costs for briquette production under the same circumstances of pelleting cost calculations. Due to lower quality requirements of the input material, (Kaltschmitt 2009) the external costs for drying and crushing can be assumed to be significantly lower with respect to pelleting. But for sake of comparability here the preliminary material treatment costs have been set equal. For local devices, the investment costs of briquetting machinery are considerably high due to developed technology and hydraulic parts.

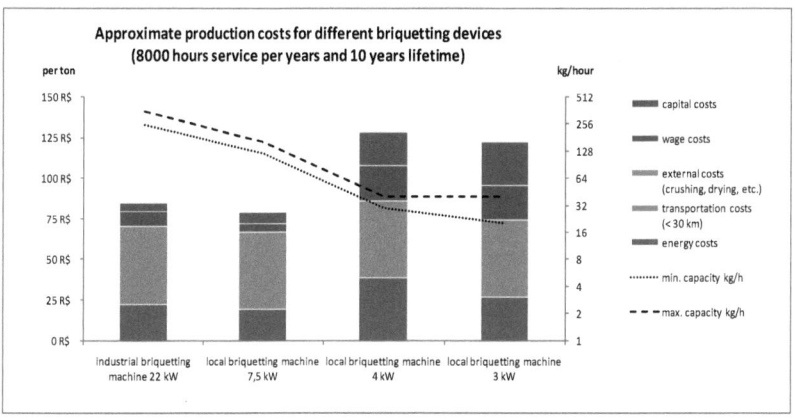

Figure 5.12: Comparison of production costs for different briquetting approaches.

As shown above, the biomass fuel production costs for small-scale machinery (power input 3 kW and 4 kW) are lower with pelleting then with briquetting devices. On this scale briquetting has as practical advantage, due to less dependency on input material quality (Kaltschmitt 2009). Therefore the desired product, a compacted biomass fuel can be produced with higher stability. Moving to medium and large-scale applications (power input > 7.5 kW), the production costs for briquettes are slightly lower. Though dependency on raw material quality still exists and costs for preliminary treatment can be significantly lower for the briquetting process it is considered as the best applicable approach.

As shown above, the biomass fuel production costs for small-scale machinery (power input 3 kW and 4 kW) are lower with pelleting then with briquetting devices. On this scale briquetting has as practical advantage, due to less dependency on input material quality (Kaltschmitt 2009). Therefore the desired product, a compacted biomass fuel can be produced with higher stability. Moving to medium and largescale applications (power input > 7.5 kW), the production costs for briquettes are slightly lower. Though dependency on raw material quality still exists and costs for preliminary treatment can be significantly lower for the briquetting process it is considered as the best applicable approach.

The compaction of biomass could be an alternative to save energy, transport and storage costs. However, due to the variety of biomass production chains in Brazil and the type of technology applied, costs vary broadly. The German experience in the biomass compaction is taken as reference to define market scenarios for biofuels in Brazil.

The fabrication of pellets and briquettes requires processing, which is dependent on primary input material. Biomass water content and particle size determine the extent of energy input. Input diameters can be from some µm (sawdust) to many cm (wood chip).

The composition of raw materials for pellet and briquette production is locally and temporally fluctuating while the process steps remain. In Brazil, briquettes are used as a fuel in several applications: industrial furnaces (pig iron), ceramic industry and bakeries. Many authors have described the environmental and socio-economic benefits of compacted biomass residues: easy handling, higher storing availability, substitutes wood logs, less emissions, higher calorific value and fuel uniformity when compared to wood.

5.3.2 Power generation case for biomass residues utilization

The power generation case represents application of unprocessed non-compacted agricultural and production residues exclusively for electricity generation. Material availability and transportation plays a major role in the cost calculations. In this case, sufficient biomass fuel supply is assumed. Transportation costs are included to biomass fuel prices. The maximum transportation distance is assumed to be 30 km.

The power generation case excludes heat-operated facilities and any profit from heat selling. Therefore, only leveled electricity costs determine cost-effectiveness of the facility. The technology for energy conversion is dependent on the prospected power plant capacity. In this survey two power plant types were chosen. Their technical frame variables are listed below.

Table 5.6 Conversion technologies for biomass combustion (adapted from Gaderer, 2009 and Schwarz, 2009)

Energy conversion technology	Electrical efficiency (%)	Thermal efficiency (%)	Initial investment cost (BRL/kW)	Typical capacity (kW_{el})

Dedicated Steam Cycle (DST)	10 – 30	80 – 90	5,300 – 8,700	5,000 – 25,000
Organic Rankine Cycle (ORC)	10 – 18	66 – 83	4,400 – 6,200	100 – 5,000
Stirling CHP[8]	9 – 12	85 – 90	8,740 – 13,570	Less than 100

5.3.2.1 Methodology for power generation case calculations

The investment costs for conversion technologies have been used and adapted to the Brazilian market, by adjusting initial prices according to comparable Brazilian reference biomass projects implemented in the Federal State of Pará by the University of Pará (Rendeiro, 2011) and in the Federal State of Ceará by the municipality Fortaleza in cooperation with the Chair of Power Plant Technology in 2009 at the Brandenburg University of Technology (BTU Cottbus).

The base investment costs refer also to the average initial investment costs given in Table 5.6. A business-planning tool used for a biomass power plant design in Fortaleza, gives all operational and investment cost factors. This was developed using template data from the Lusatian biomass power plant in Sellessen (Vattenfall, 2011). The cost calculations for different fuel materials and technologies were performed with a broad data input.

Electricity output accounts as exclusive return of the biomass plant. Application of heat and CO_2 certificates were excluded in the primary step. The power plant capacity was set with 200 kW_{el} due to small-scale approach and comparability to a case study in Pará (Rendeiro, 2011). Design referring to annual load curve does not apply for electricity-operated facilities. According to Table 5.6, the electrical efficiency of a power plant is dependent on the capacity. In order to achieve a practical comparison, the efficiency of a DST plant was set to maximum and for ORC plant set to minimum. The reason for such comparison is that the efficiency range from the ORC technolgy is relative small. On the other hand, the efficiency range from DST technology is larger (Table 5.6). This enables to present a range of possible costs.

Many cost factors are sufficiently described with linear dependency on power plant size (investment, operational and other costs). As mentioned above, the biomass power plant

---

[8] Biedermann (2004)

in Sellesen and additionally a related business plan for Brazil (Fortaleza) have been used as cost factor basis. By extrapolation to other power plant capacities the cost factors were adapted. The behavior of personal costs is not linear. Here approximations from practical experience were used to estimate working effort (personal communication).

The time frame for amortization of the reference power plant was set to 10 years. This period refers also to the minimum service lifetime. Linear depreciation has been excluded by aiming at cost-benefit comparison before tax. In case of extension of the service life over amortization period, the annual capital costs only include the area rent, because of no further debt. Therefore it is called service life, which is the period of payback for any initial investment for facility construction. In the Figure below, the annuity of a reference power plant using ORC technology was used for the sake of simplicity, in order to illustrate the behaviour of annual capital costs.

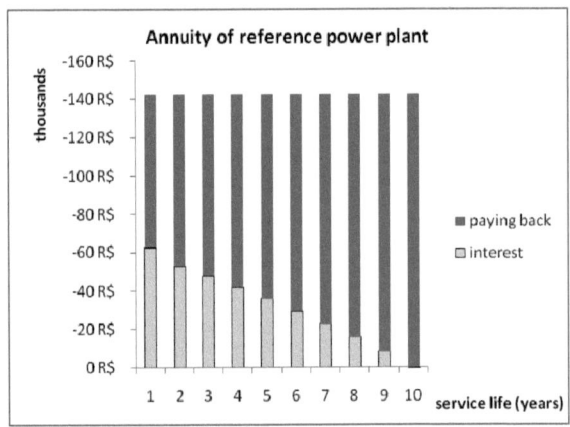

Figure 5.13: Example payback calculation of an ORC power plant (200 kW$_{el}$)

The running costs of a biomass power plant are distinguished in fuel costs and operation costs. The latter includes all involved other material costs like cooling water, disposal cost, fuel for feeding devices and internal electricity usage and is derived by extrapolation from the reference case of a biomass power plant in Selessen (Vattenfall, 2011). For electricity consumption of the power plant an average price has been applied, in order to represent annual and regional fluctuations. Table 5.7 gives complete information which data has been used for the economic feasibility of the electricity generation case calculations.

A list of present electricity market prices was used as a base for the return calculation (ANEEL, 2011). By using alternative customer prices, electricity market competition can be examined and economic feasibility determined. According to the Brazilian legislation, the power plants are obliged to sell all electricity produced for the regional concessionary. Therefore the generated electricity is sold to the regional concessionaries. For the return calculations, it was assumed that the selling price for produced electricity should be equal or lower to the tariff charged from the regional concessionary in order to achieve feasibility.

When applying the business-plan for large-scale power plants, electricity prices should be considered according to the energy auctions opened by the Brazilian Ministry of Mining and Energy (see Chapter 2). In the case of Minas Gerais, the CEMIG[9] is responsible and in the case of Pará, the CELPA[10]. For biomass the least auction was scheduled in 2008 with 155 BRL / MWh as nominal leveled energy cost to undercut. The large-scale approach doesn't have to be considered for the prospected nominal capacity of 200 kW_{el}.

Table 5.7: Cost factors for calculating the electricity generation case

---

[9] CEMIG: Companhia Energética de Minas Gerais
[10] CELPA: Centrais Elétricas do Pará S.A

Cost factor	Unit	Value	Data Background
General Data			
Maintenance cost	%	3	Often applied value
Disposal - cost for ashes	per t	100	PC[11]
Annual field rent	per year	16,700	Template
Fuel derived data			
Heating oil demand (extra power device)	l/a	0	Extrapolation
Drinking water	per m³	5.85	Extrapolation
Well water	per m³	0.002	Arbitrary
Ash per lorry	T	10	Common capacity
Price/ash transport	per transport	150	CAGECE[12] list price
Rent for toploader	per hour	20.3	Template
Waste water price	per m³	0	PC
Well water treatment	per m³	0	PC
Internal Electricity data			
Peak load fee (12 h)	kW / month	23.18	CELESC[13] list price
Base load fee (9 h)	kW / month	4.3	CELESC list price
Average connection	Hours / day	12	Arbitrary peak load time
Peak load prices (consumption)			
Dry season	per kWh	0.25434	CELESC list price
Wet season	per kWh	0.22996	CELESC list price
Base load prices (consumption)			
Dry season	per kWh	0.15834	CELESC list price
Wet season	per kWh	0.14419	CELESC list price
Other costs			
Insurance (2,5 % of invest cost)	per year	2.5%	Template
Demolition (0,0055 % of inv.)	per year	0.00055	Template

The consumption of external electricity for power plant operation can contribute a large amount to annual running costs. Therefore two scenarios have been applied. The **external electricity usage** scenario assumes complete selling of generated electricity

---

11 Personal communication
12 Companhia de Água e Esgoto do Ceará
13 Electricity concessionary in Santa Catarina

and purchase of electricity with prices charged by the regional concessionary CELESC. The **internal electricity usage** scenario assumes coverage of internal electricity requirements by own generation. Hence the latter approach enables grid-independent operation of the power plant, when objecting island solutions. Also this is an option to lower running costs when internal generation is cheaper than external purchasing.

For the initial financing of power plant implementation two different approaches apply. They represent viewpoints of the facility owner. The **self-financed approach** assumes full payment of the facility by the owner, by means of credits without any preliminary owned capital. Here the service life period is used to payback creditors from facility returns. With the annuity method, required annual payments can be calculated. The applied interest rate (6 % p.a.) was taken from the business plan for Fortaleza. Using an annuity credit, annual payments are constant for the duration of service life (see Figure 5.13). The resulting annuity (paying back plus interest) accounts to a large share for capital cost of the facility. The **"grand fathering" approach** assumes donation of the reference power plant by government or any other entity. Here only area rent accounts to capital cost, which are reduced to a minimum.

Grid service fees are neglected as they are considered to be included in connection costs (Table 5.7). The annual power plant revenue before tax is calculated as difference of returns minus all costs. By weighting all costs with annual electricity output, the specific generation costs (leveled energy cost) can be determined. Both results are comparable with existing references.

5.3.2.2 Results

By determination of all cost factors and returns from electricity selling in Minas Gerais and Pará, the economical situation of two applicable technologies can be described. The strong dependency on biomass fuel prices can be observed in a cost function. For Minas Gerais, the feasibility of biomass power plants was calculated for sugar cane bagasse, coffee husks, rice husks and eucalyptus saw wood. For Pará, sawdust from native woods was considered in the calculations.

In Figure 5.14, the annual return and annual material costs for the **self-financed approach** in Minas Gerais with **external electricity usage** are presented comparing sugar cane bagasse (w 50 %) and coffee husks. Since sugar cane bagasse (w 50 %) is available free of charge, transportation costs for local applications are reduced to zero. Transportation costs account (100 BRL/ton), if the local bagasse supply is not sufficient for fuel input. This case is referred to bagasse (w 12%), which already is dried for transportation over a distance less than 30 km.

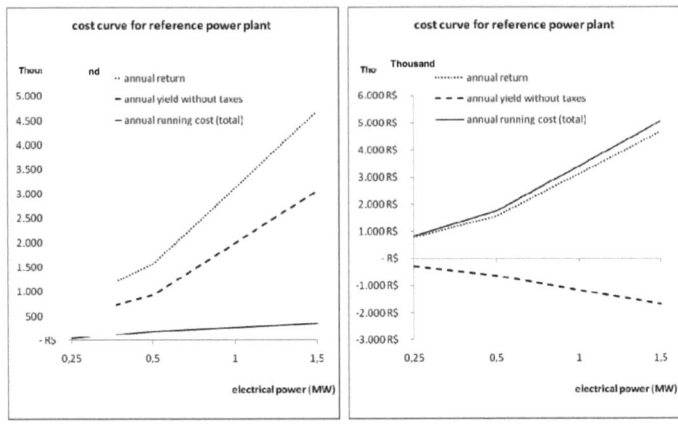

Figure 5.14: Economic feasibility for biomass plants using sugar cane bagasse (w 50 %) (left) and coffee husks (right), considering the self-financed approach

The first year returns account as revenues of electricty selling. Generally, in economic balancing the first year is considered as construction and commissioning period. Therefore it is assumed that the power plant operates constantly and faultless from the 1st day of functioning, which is similar to the 1st day of economical balancing. For coffee husk application in Minas Gerais, the return hardly covers expenses on material, so the revenue before tax for coffee husk application is negative. For bagasse (w 50 %), which is available free of charge, the return exceeds material cost significantly. Assuming sufficiently available biomass, the investment in a biomass power plant of the given scales is beneficial.

For coffee husks, a market price (60 BRL/ton) has to be paid additionally to transportation costs. This raw material has been chosen to represent biomass residues, which already have an alternative market (chicken farms). The economic feasibility of eucalyptus sawdust and rice husks is similar to coffee husks, due to slightly deviating physico-chemical parameters and comparable process chains.

As shown in the Figure 5.14, the utilization of coffee husks, or other type of biomass residue with similar supply chain for power generation is not feasible when under the following conditions (the ORC technology was used for illustrating a possible worst case scenario):

- self-financed investment (6% p.a. interest rate)
- ORC technology (10 % electrical efficiency)
- located in Minas Gerais
- CEMIG has commercialized energy at an average cost of BRL 390 / MWh

In order to examine feasibility of electricity supply, depending on financial support, the self-financed approach and the "grand fathering" approach were compared. Although "grand fathering" lowers annual capital costs drastically, positive revenue for coffee husk and ORC technology with minimum efficiency cannot be achieved in Minas Gerais. Since coffee husks, eucalyptus sawdust and rice husks have similar supply chains and properties, a power plant using this biofuels could not be economically feasible under the above listed conditions.

In Figure 5.15, sugar cane bagasse is presented as potential biofuel for such conditions. Hence, revenues for both financial approaches could be analyzed. Figures 5.15 illustrates that the revenue with "grand fathering" is higher than for self-financed approach, due to the minimized capital costs. Although self-financed projects are feasible for sugar cane, "grand fathering" through governmental support or international donations is an attractive option.

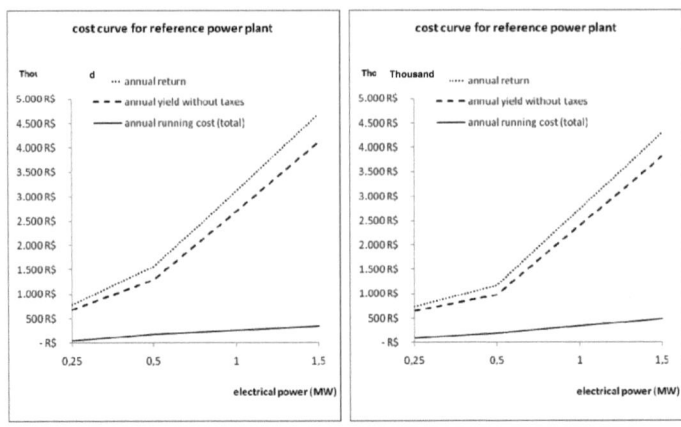

Figure 5.15: Economic feasibility for sugar cane bagasse plant, considering grand fathering approach and external electricity usage (left), internal electricity usage (right)

By using specific electricity generation costs, a more detailed description of feasibility is possible. In Brazil electricity prices vary regionally which can trigger feasibility in one region and hinder in another. The range of prices is very large. Therefore regional specification for a biomass power plant project is necessary. By using minimum efficiency for ORC and maximum for DST, feasibility borders can be determined. The average costs from internal and external electricity usage were used to calculate the electricity costs of a 200 kW power plant, considering different financing scenarios and technologies.

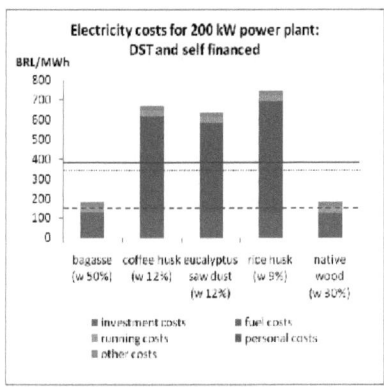
(a)

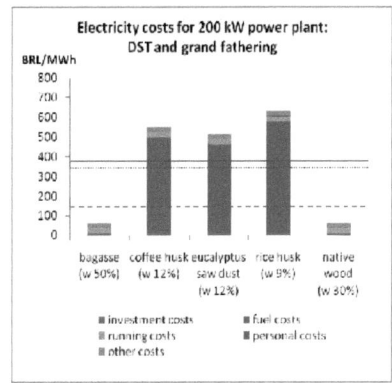
(b)

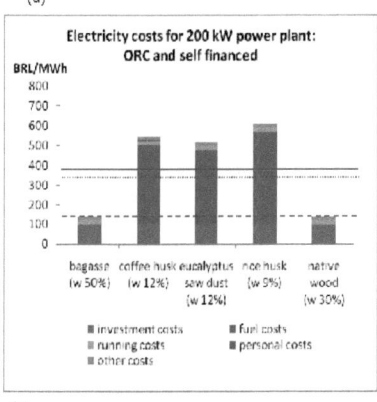
(c)

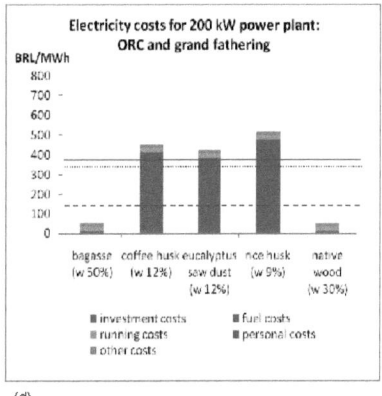
(d)

Figure 5.16: Electricity generation costs for 200 kW$_{el}$ nominal power plant under Brazilian price conditions: CEMIG (black line), CELPA (grey line) and auction scheme (dashed line) and different schenarios: (a) DST and self-financed; (b) DST and grand fathering; (c) ORC and self-financed and (d) ORC and grand fathering

The specific electricity generation costs for bagasse (w 50 %) are considerably low. Due to the minimum fuel price here 5 BRL / MWh can be achieved at the lower border for "grand fathering" and 16 BRL / MWh for self-financed approach. This allows external electricity usage in any case and participation in auction schemes.

The achieved return from electricity selling (external el. usage) still compensates expenses on electricity purchasing, assuming an electricity consumption price of 190 BRL / MWh, including connection cost. For island solutions (internal el. usage), according to the above listed conditions, the possible return is lowered by 6 % in average (see Figure 5.15) due to required internal electricity usage. For the considered capacity (200 kW$_{el}$) annual internal electricity consumption accounts 4% of generation.

A combination of technically possible power plant efficiencies (Table 5.7) has been analysed step-by-step. The aim was to achieve a cost coverage of a power plant in operation, with 200 kW$_{el}$ nominal power. By using different electrical efficiencies, dependent on applied technology and input raw material, the economic feasibility was determined in Table 5.9.

Table 5.8: Economic feasibility for different financing approaches, technologies and biomass

Raw material	Organic Rankine Cycle	Dedicated Steam Cycle
Grand fathering	Economic Feasibility	
Bagasse (w 12%)	high	high
Bagasse (w 50%)	high	high
Eucalyptus sawdust (w 12%)	high	high
Coffee husks (w 12%)	high	high
Rice husks (w 9%)	low	high
Native wood (w 30%)	high	high
Self financed	Economic Feasibility	
Bagasse (w 12%)	low	high
Bagasse (w 50%)	high	high
Eucalyptus sawdust (w 12%)	low	low
Coffee husks (w 12%)	no	low
Rice husk (w 9%)	no	no
Native wood (w 30%)	high	high

The raw sugar cane bagasse with water content of 50 % is applicable in any mode if the material and transportation price is zero. On the other hand, the utilization of coffee husks, eucalyptus sawdust and rice husks for power generation, considering the investigated supply chains in Minas Gerais, could not be feasible for self-financed biomass projects using ORC technology.

Due to slightly higher investment costs (see Table 5.6) for DST when compared to ORC, auction prices cannot be undercut. But alternative electricity supply for remote areas, with island solutions is still beneficial. Raw materials with an overall price above R$ 160 can be considered as hardly applicable to any mode of selling.

5.3.3 Integrated case for biomass residues utilization

The integrated case has as main compensation effect, the biomass compaction, which allows selling of an additional product. Additionally material transportation over longer distances and storage abilities are improved. The main objective of the cost examination is to calculate feasibility scenarios for combined heat power generation with the additional production of a marketable biofuel. The integrated concept refers to an industrial cluster. Here agricultural and production residues serve as fuel for a CHP facility. Additionally the excess amount of material is processed into briquettes for commercialization.

The integration of waste heat for preliminary material drying is a technically feasible process step (B&B Bioenergie, personal communication). According to the facility located in Calau, Germany, the required drying temperature is considerably low, about 50 °C. The theoretically required heat annual amount for hot air drying with the belt dryer, assuming low dryer efficiency with 30 % is calculated with 370 kWh_{th}/ton. This thermal energy can be provided by waste heat of the power plant. The additional equipment with heat exchangers and drying chambers increases initial investments cost, but the amount is comparatively low with respect to combustion technology. Electricity consumption cost of the dryer (belt dryer in Calau) is provided by the power plant and included to the balance as internal electricity requirements.

Compared with the calculated annual heat amount of approximatelly 7,300 MWh_{th}/year in a power plant with 200 kW_{el} nominal power (ORC with 18% efficiency), sufficient heat can be provided for drying. Therefore the water content of bagasse can be reduced from 50 % to 12 mass %. This enables to exploit a higher calorific value parallel to zero transportation cost, due to local availability of bagasse (w 50%). By reducing the raw material price to zero, fuel costs do not contribute to leveled energy costs. Hence the amount of applied fuel is regardless for price calculations and therefore electrical efficiency no longer affects economic feasibility of the integrated power plant. Lower electrical efficiency requires a larger amount of fuel. This has a back-effect on drying because larger raw material input requires more drying energy. Based on annual fuel input and available thermal energy the feasibility of integrated drying is calculated.

For different electrical efficiencies different amounts of fuel are required. Due to regional organization of biomass supply the availability is limited. In order to check sufficient supply of biomass the required raw material amount for a 200 kW$_{el}$ facility is calculated. Primarily the analysis of the integrated case examines local electricity generation without raw material transportation. Therefore bagasse and native wood have been chosen as fuels in the first step. Due to their availability as free of charge raw material they remain locally at their generation source. The technical frame conditions concerning cost factor extrapolation, service lifetime and service hours remain.

The annually produced amount of sugar cane bagasse was reported with 200 ton, in the prospected region. The required amount (4,100 ton / year) exceeds the locally available material by far. Therefore additional bagasse has to be transported to the power plant in order to operate with 200 kW$_{el}$. No excess raw material is available, which disables the option of material compaction. With a local fuel supply approach a power plant with 200 kW$_{el}$ has too high fuel requirements.

A feasible scenario for an integrated power plant, without additional fuel transportation to the investigated community, would provide a nominal electrical power below 100 kW. This rather micro-scale of generation requires an applicable technology. The above listed ORC and DST processes are not suitable. Therefore a Stirling CHP facility has been chosen as reported by Obernberger (Biedermann, 2004). Here the waste heat can be used for bagasse drying and integrated to the local production of sugar.

As resource integration bagasse (w 50%) from sugar cane refining can be handed into a local dryer, driven by waste heat from the power plant. The dried bagasse will be used as fuel for the Stirling CHP facility. The generated heat is used primarily used for Stirling operation, secondary in the sugar refining and tertiary in the bagasse drying process. The generated electricity is used internally for machine operation (internal electricity usage) and the excess amount can be sold to the local concessionary.

Assuming combustion of internally dried bagasse with water content of 12 mass % fuel input to the power plant can be calculated. After evaporation of water from the raw bagasse (w 50%) approximately 114 ton bagasse (w 12%) per year can be used as fuel. The drying of bagasse requires heat energy input. By applying hot air drying with 30% efficiency, the required annual amount of heat is calculated with 74 MWh$_{th}$/year (B&B Bioenergie, personal communication). The electrical energy consumption of 5

MWh$_{el}$/year by the dryer is considered as internal electricity usage. The investment costs of drying technology (1,000 BRL / year) are included in the capital costs. The facility service lifetime is fixed to 10 years, assuming 8,000 annual working hours.

Table 5.9: Economic analysis of a 7 kW$_{el}$ sugar cane bagasse CHP plant with Stirling technology

Small-scale sugar cane bagasse CHP plant in Minas Gerais		
	Self-financed	Grand fathering
Specific electricity generation costs	BRL 41 MWh	BRL 23 MWh
Capital cost	BRL 11,200 / year	BRL 2,700 / year
Planning cost	BRL 0 / year	BRL 0 / year
Running cost (fuel)	BRL 0 / year	BRL 0 / year
Running cost (material)	BRL 2,100 /year	BRL 2,100 /year
Running cost (personal)	BRL 3,900 /year	BRL 3,900 /year
Insurance and other fees	BRL 1,700 / year	BRL 1,700 / year
Return	BRL 15,800 / year	BRL 15,800 / year
Annual yield without taxes	-BRL 3,100 / year	BRL 5,400 / year
Annual electricity generation	43 MWh / year	43 MWh / year
Initial investment cost	BRL 63,0	BRL 63,0

As indicated in Table 5.9, this approach cannot achieve cost coverage for a self-financed approach. For such micro-scale solutions even optimization of small cost factors, like field rent can result in better feasibility. In order to find a feasible solution the breakeven point was calculated for integrated drying of bagasse in a local micro-scale power plant.

For the self-financed approach this facility requires at minimum 11 kW$_{el}$ nominal power in order to achieve cost coverage. Here the leveled energy costs equal regional electricity customer prices charged by CEMIG (390 BRL/MWh). The returns from electricity selling cover generation costs. Due to calculation model deviation (+ 10 %) this value should not be handled as an absolute threshold, but as a feasibility range.

Considering the grand fathering approach, the use of bagasse for heat and power generation combined with compaction is feasible at cost coverage with a 4 kW$_{el}$ facility.

With self-financed approach, the power plant capacity would be 9 kW_{el}. Both capacities implies sufficent bagasse supply. The lower border of economic feasibility is between 4 and 9 kW_{el} nominal power. The required area for the production of 16 tons bagasse was indicated with 1 ha. Therefore production area is between 5 ha and 15 ha in order to get sufficient supply from a regional source. These acreage requirements are still on local scale.

It is imporatnt to mention that the Stirling CHP facility is a cutting edge development, which not yet has achieved broad availability (Biedermann, 2004). The integration of a dryer to a sugar cane bagasse facility could be a development objective.

For the application of native wood in the Federal State of Pará, the same integration concept can be applied for drying. Due to the lower raw material water content less drying heat is required. Additionally excess amounts can be used for biomass compaction and product selling. A broad range of excess amount for native wood can be achieved by applying different electrical efficiencies of conversion technology. The proposed compaction technology is briquetting because less raw material preparation is required and parallel good product stability can be achieved with local machinery operation staff (Schuco 2011, personal communication). Capacity building for briquetting is required and feasible. Energy costs and external costs for the briquetting machinery, which result in electricity costs, are indicated in the power plant balance (internal electricity usage).

The return loss from electricity selling is calculated for the power plant and additional return from wood briquette selling is included. The whole approach is presented as island solution due to remote localization in Pará. As discussed above the nominal efficiency is decisive for material input. Here the ORC technology was chosen with maximum electrical efficiency (18 %) in order to present a real value based on the investigated case study in Pará (Rendeiro, 2010). The nominal power has been set to 200 kW_{el}, in order to achieve comparability with the investigated case in Pará (see Chapter 6).

The native wood sawdust originated in local sawmills is used as fuel. The preliminary drying of sawdust is done with waste heat from the power plant. A briquetting device performs the compaction of surplus raw material, with electricity supply from the power plant. Assuming 200 kW_{el} installed power; the required raw material of native wood (w

30%) is 2,800 ton / year. This results in an excess amount of 1,740 ton / year of native wood, which can be processed to briquettes. Therefore a briquetting machinery with nominal power input of 7,5 kW and 350 kg/h maximum capacity has been chosen. Two of these devices are required in order to produce pellets from the calculated excess sawdust. The machines are assumed to be operating 8,000 hours per year with power supply from the power plant. The consumed electricity is included in the power plant balance as internal electricity usage.

Briquette selling in Brazil with the nominal market price of 300 BRL / ton (Felfi et al., 2010) increases entrepreneurial return. The loss from decreased electricity selling increases leveled energy cost. The additional investment and wage cost as well as lost returns from electricity selling can be compensated, as shown in Table 5.10: the calculation assumes availability of the state of the art technology for ORC process, drying and briquetting machinery.

Table 5.10: Economic analysis of a self-financed 200 kW_{el} CHP plant with ORC technology using native wood sawdust in Pará

	Electricity generation native wood (w 30%)	Drying native wood (w 12%)	Briquette production native wood (w 12%)
Spec. elec. generation costs	137 BRL / MWh	129 BRL / MWh	155 BRL / MWh
Capital cost	142,000 BRL/year	148,000 BRL/year	152,000 BRL/year
Planning cost	0	0	0
Running cost (fuel)	0	0	0
Running cost (material)	33,000 BRL/year	35,000 BRL/year	36,000 BRL/year
Running cost (personal)	8,100 BRL/year	10,000 BRL/year	16,000 BRL/year
Insurance and other fees	29,000 BRL/year	29,000 BRL/year	29,000 BRL/year
Return	555,000 BRL/year	550,000 BRL/year	1,026,000 BRL/year
Annual yield without taxes	343,000 BRL/year	328,000 BRL/year	793,000 BRL/year
Annual electricity	1,550 MWh/year	1,485 MWh/year	1,320 MWh/year

generation

As shown in Table 5.10:, the integration of biofuel production is an additional income and employment source. Higher investment and running costs are compensated through returns from briquette selling. For the investigated power plant (200 kW$_{el}$ ORC) a total investment cost of approximately 1.2 million BRL is necessary. The specific investment cost is 5,300 BRL / kW$_{el}$. The amount of annually generated electricity is 1,300 MWh/year with integrated briquette production. Additionally a renewable energy product is generated.

The margin of the whole production process is strongly dependent on the raw material price of native wood. By creating a market for this residue, further price increase is possible. Based on the calculations presented in Table 5.10: the raw material price inclusively transport cost should not exceed 105 BRL / ton. Here break even is achieved for the integrated production of briquettes from native wood. At this point return from electricity and briquette selling still covers all expenses on production.

The production of bagasse briquettes could be also implemented in a 200 kW$_{el}$ facility (ORC with electrical efficiency of 18 %) in other Brazilian regions. The resulting marginal costs for bagasse are higher than 130 BRL / ton. This could be regarded as the maximum raw material price using only bagasse as fuel for an integrated process with additional briquette selling. The production is considered cost effective when the price of the raw material is lower than 130 BRL / ton. However, the application of coffee husk, eucalyptus saw dust and rice husk for electricity generation, in Minas Gerais has been shown as inefficient. These residues have already a supply chain and a market price. Moreover, the quantity available is not sufficient for operating a 200 kWel power plant using ORC and DST technologies.

5.4 Discussion

Biomass costs of dedicated production systems are especially dependent on the availability of residues, transportation costs, availability of human labor and existing supply chain. Through investigating several biomass energy generation schemes, the author could conclude that in general, there is a technical and economical feasibility for residues and wastes.

Heat and power generation plus the compaction of biomass residues was not identified in Minas Gerais. On the other hand, biomass projects and the commercialization of briquettes is giving the first steps in North Brazil. The author visited a briquette facility in the Federal State of Pará, city of Belém. The huge amount of wood residues produced by sawmills in this region and its low price in the market, could allow North Brazilian briquettes competing with the European market.

Nevertheless, in North Brazil, wood residues are mostly originated from illegal timber falling. As this wood production is not certified, it would be not possible to produce and export briquettes and pellets with residues of this production (Walter and Dolzan, 2009). Nevertheless, the author investigated 3 sawmills in the Federal State of Pará (Amazon) where timber was certified by the government.

As reported by the author, most of the sawdust generated in Southeast Brazil is sold to chicken farms, to be used in chicken beds. Regarding the compaction of the residues into briquettes and pellets, there is no precise estimation of the amount of production in the country using sawdust, wood chips, wood waste and scrap. Since already exists the commercialization of wood residues, introducing new technologies and a new market for pellets or briquettes for local use or for exporting is still a challenge for Brazil.

The carpentry industry generates most of the residues, processing the entire wood logs for the production of timber pilling and/or oriented strand boards (OSB) or chipboards. During the first steps of manufacturing – wood peeling and sawing – there is the production of great quantities of shavings, which could be used for energy generation in loco. On the other hand, the enterprises which process wood for the manufacture of noble products, like furniture, tend to produce less amounts of residues and with smaller particle size, like saw wood and saw powder. These materials could be compacted in loco, diminishing the storage and transportation constraints.

5.5 References

ANEEL - Agência Nacional de Energia Elétrica (2011) Tarifas residenciais. Available at: http://www.aneel.gov.br/area.cfm?idArea=493&idPerfil=4 (Last acessed on Mai 27, 2011)

B&B Bioenergie (2011) Personal communication. Available at: www.bioenergie-calau.de/

Benedetti M., Fugiwara A., Factori M., Costa C., Meirelles P. (2009) Adubação com cama de frango em pastagem. Águas de lindóia. Zootec. cd rom.

Brauer (2009) Elektrotechnik Brauer. Available at: http://www.elektrotechnik-brauer.de/images/bhkw/jahresdauerlinie_big.jpg (Last acessed on Mai 27, 2011)

Biedermann F., Carlsen H., Schöch M. and Obernberger I. (2004) Available at: http://www.bios-bioenergy.at/de/downloads-publikationen/kwk-stirlingmotor.html Last acessed on Mai 27, 2011

Dena (2009) CO_2 Bilanzierungsstudie 2009. Zuarbeiten der Universität Göttingen Available at: www.uni-goettingen.de/de/79037.html Last acessed on Mai 27, 2011

Desplechin E. (2011) Brazilian bioelectricity from bagasse. Available at : http://english.unica.com.br/opiniao/show.asp?msgCode={73BF1846-A8CC-4494-943C-FC2BAE4B1C06}(Last access 6 April 2011)

Eichhorn H. (1999) Landtechnik: Landwirtschaftliches Lehrbuch. Stuttgart : Ulmer, 1999.

EPE - Empresa de Pesquisa Energetica (2008) Available at: http://www.epe.gov.br/Paginas/default.aspx (Last access 21 June 2011)

EPE - Empresa de Pesquisa Energetica (2010) Available at: http://www.epe.gov.br/Paginas/default.aspx (Last access 21 June 2011)

FNR - Fachagentur für Nachwachsende Rohstoffe (2001) Energetische Nutzung von Stroh, Ganzpflanzengetreide und weiterer halmgutartiger Biomasse. Tautenhain

FNR - Fachagentur für Nachwachsende Rohstoffe (2007) Handbuch Bioenergie Kleinanlagen. Gülzow

Frota WF. and Rocha BRP. (2010) Benefits of natural gas introduction in the nergy matrix of isolated electrical system in the city of Manaus – sate of Amazonas – Brazil. In: Energy policy 38: 1811-1818

Gaderer M. (2009) Organic Rankine Cycle. Available at: http://www.carmen-ev.de/dt/portrait/sonstiges/gaderer.pdf Last accessed on 9 January 2010.

GEMIS (2009) Ökoinstitut Deutschland e.V. Available at: http://www.oeko.de/service/gemis/en/index.htm Last accessed on 9 January 2010.

Gladstone N. (2009) Legacy Foundation. Actionsheet 77. Available at: www.paceproject.net. Last accessed on 9 January 2010.

Golser M., Pichler W. and Hader F. (2010) Energieholztrocknung Kooperationsabkommen Forst-Platte-Papier, Österreich, Wien, 2010. Available at: http://www.klimaaktiv.at/article/articleview/43383/1/13846/ Last accessed on 9 January 2010.

Higashikawa FS., Silva CA. and Bettiol W. (2010) Chemical and physical properties of organic residues. In: R. Bras. Ci. Solo 34: 1743 - 1752

IEA - International Energy Agency (2007) IEA Energy Technology Essentials. Available at: http://www.iea.org/Textbase/techno/essentials.htm

INCOFEX (2011) Personal communication

Jamieson S. (2007) Canadian Forest Industries Magazine. Saving in Scandinavia. Available at: http://canadianforestindustries.ca/app/newsletter/view_article/31,2,1,,1.html Last accessed on 9 January 2010.

Jossé G. (2007) Basiswissen Kostenrechnung. s.l. : Deutscher Taschenbuch Verlag, 2007.

Roßmann J. (2009) A Detailed Timber Harvest Simulator Coupled with 3-D Visualization. World Academy of Science, Engineering and Technology 57, S. 243 - 248.

Kaltschmitt M., Hartmann H. and Hofbaier H. (2009) Energie aus Biomasse; Grundlagen, Techniken und Verfahren. heidelberg : Springer, 2009.

Kraft (2011) Personal communication. Fa. BauerPower Geschäftsführer, kraft@energievomland.de

Kreimes H. (2008) Den Wald „verdoppeln". Heizwerterhöhung durch Hackschnitzel-Trocknung in einfachen Anlagen. Available at: http://www.carmen-ev.de/dt/portrait/sonstiges/biokraftstoffkongress08/05_Kreimes_FH_Rosenheim.pdf. Last accessed on 9 January 2010.

Mansaray KG. and Ghaly AE. (1997) Physical and thermochemical properties of rice husks. In: Energy Sources 19, Issue 9. Taylor and Francis

Olfert K. (2008) Kostenrechnung. s.l. : Kiehl, 2008.

Pereira MG., Freitas MA. and Silva NF. (2010) Rural electrification and energy poverty: empirical evidences from Brazil. In: Renewable and Sustainable Energy Reviews 14 (4) pp. 1229-1240

Pereira MG., Freitas MA. and Silva NF. (2011) The challenge of energy poverty: Brazilian case study. In Energy Policy 39. Elsevier pp. 167-175

Quaschning V. (2009) Erneuerbare Energiesysteme. München : Carl Hanser Verlag, 2009.

Rendeiro G. (2010) Geração de energia elétrica em localidades isoladas na amazônia utilizando biomassa como recurso energético. Tese de doutorado. Programa de Pós-Graduação em Engenharia de Recursos Naturais da Amazônia - PRODERNA. Instituto de Tecnologia.Universidade Federal do Pará. Agosto 2011.

Schuco (2011) Personal Communication

Schwarz and Rosanelli (2009) ORC Technologie. Available at : http://www.provinz.bz.it/landwirtschaft/download/5._ORC-Technologie.pdf. Last assessed on 9 January 2010.

Saenger M., Hartge EU., Werther J., Ogada T. and Siagi Z. (2001) Combustion of coffee husks. In: Renewable Energy 23: 103-121. Elsevier

Seipp (2011) Personal communication. Fa. Seipp Maschinenring zur mobilen Pelletierung LU-Seipp@t-online.de

Steinborn, E. 2009. Wirtschaftlichkeit von Blockheizkraftwerken. Available at: http://www.bhkw-info.de/wirtschaftlichkeit/bhkw_wirtschaftlichkeit.html Last assessed on 30 June 2011

Vatenfall (2011) Kraftwerk Selessen. Available at: http://www.vattenfall.de/de/new-energy-erzeugungsanlagen.htm Last assessed on 30 June 2011

Walter A., Dolzan P., Quilodrán O., Garcia J., da Silva C., Piacente F., Segerstedt A. (2008) A Sustainability Analysis of the Brazilian Bio-ethanol. Unicamp. Department for Environment, Food and Rural Affairs (DEFRA), British Embassy, Brasilia.

Walter A. and Dolzan P. (2009) Brazil country report – 2009. In: Country Report: Brazil – Task 40 – Sustainable Bio-energy Trade; securing Supply and Demand. Available at: http://www.bioenergytrade.org/downloads/brazilcountryreporttask40.pdf (Last access 16 April 2011)

Weima (2011) Personal communication Fa. Weima Brikettierungsanlagen, Vertriebsgebiet Neue Bundesländer

Wornath (2011) Personal communication. Fa. RUF Brikettierungsanlagen, Vertriebsgebiet Lateinamerika

6 Decentralized biofuel schemes in Brazil

The success of the large-scale decentralized bioenergy schemes in Brazil is due to the sugar cane industry. Ethanol production demands a high amount of energy input. Therefore, the combustion of bagasse became an alternative for decking the energy demand in the ethanol plants and supplying the National Grid with bagasse electricity in the dry season, when the shortages in hydropower often occur.

Even when positive contributions of, for instance, the sugarcane and ethanol industry on job creation and income in Brazil have been recognized; in its agricultural sector, the level of social benefits and child labour are still important issues. Several authors (Azanha, 2007; Bartocci, 2009; Cardoso, 2010 and Gomes, 2010) often described socio-economic implications of large-scale energy plantations. Data concerning social welfare, quality of employment, potential use of child labour, education and access to health care in the sugar cane fields are nowadays available. Violation of land property rights of small farmers, or the additional pressure on land and on valuable ecosystems such as rain forests are also concerning aspects (Junginger et al. 2006).

Despite of the level of formalization of the sugar cane sector is higher compared to other monocultures, the working conditions have been criticized. The evolution of mechanical harvesting in unburned sugar cane fields is already occurring, motivated by new specific legislation and environmental pressures. The ethanol industry employs approximately 1 million workers (Shikida, 2010). If mechanical harvesting would be adopted abruptly, it could cause social problems due to loss of thousands of jobs in rural areas (Hassuani et al. 2005).

The sustainability of energy from biomass resources is an essential aspect for its consolidation in the local, national and international markets (Walter et al. 2008). The generation of bioenergy will be only justified if their socio-economic and environmental impacts are favorable, compared to conventional energy sources.

Brazil is a rich country with an extreme unequal distribution of resources and income. Education, health, food and electricity are scarcely available in remote areas. Most of the families (90 %) who do not have access to electricity at home have an income equivalent to less than three minimum wages. Around 33 % of these earn less than one minimum wage[14]. The exploitation of human labour in the agro-business fields and the absence of electricity in remote areas hinder socio-economic development. Poverty and lack of perspective cause migratory flows from the countryside to the Brazilian capitals. The question that arises is how to keep farmers and rural inhabitants in their land.

In the previous chapters, the properties of biomass residues, the diversity of national approaches to electricity market reform and the economic feasibility of biomass power plants were investigated. The results show that coffee husks, rices husks, sugar cane bagasse and wood residues could be technically and economically employed as potential biofuels, only when considering the regionality and the existing supply chains of the investigated biomass. Therefore, the great challenge is how to address biomass energy generation issues to small-scale farmers and the rural population in different regions.

In order to approach the possibilities of biomass use in small-scale decentralized schemes, it was conducted a structured analysis and several cases were investigated in different Brazilian regions. Regarding data, they were collected in the Federal States: Minas Gerais (four cases) and Pará (two cases). The case studies were sorted according to the availability of electricity in rural areas. In Minas Gerais, the National Grid (hydropower) supplies all the biomass case studies. In Pará, the isolated grid system predominates and in all case studies electricity was provided through small-scale decentralized bioenergy schemes.

6.1 Approaching smallholders' views and concerns

During the research, the author observed that a high rate of illiteracy was present in the visited rural areas. Despite of the lack of income and electricity in some communities, television is the main source of information for most inhabitants. Most families are used to watch mainly soap operas. In Brazil, there is a social accommodation caused by mass media consumption, which hinder rural populations to acquire the knowledge and skills

---

[14] Brazilian minimum wage: BRL 545.00 (March, 2011) http://www.portalbrasil.net/salariominimo.htm

that could lead to emancipation. In addition, as Barros (1994) concluded, even the literate in Brazil does not cultivate a habit of reading books or newspapers. Within this context, family, neighbourhood, school and more recently television have been forming the cultural, environmental and social structures of the Brazilian rural society.

The investigation was done through interviews and literature research. The author spoke with academics, non-profit-organizations, community leaders and farmers. This provided valuable information about the chain of biomass utilization, the application of residues and the past electrification efforts in a particular region. Informal talks were used to identify the cultural background of people, the availability of resources, the socio-economic situation, the ecological and political limitations of the investigated biomass schemes. The results were interpreted and sorted according to previous research methods performed by Zerriffi (2007). This research method employs a case-by-case analysis, which was adapted according to crucial issues of energy generation schemes, like:

- Electricity supplier: centralized (hydropower) or isolated?
- Current application: heat or/and power?
- Stakeholders (beneficiaries): households, enterprises or community structures?
- Financing: how it was obtained and what does it cover?

The author selected five types of biomass after investigating the production and supply chains of residues. The description is shown above.

Table 6.1: Screening of 5 biomass utilization schemes in six different case studies in Brazil (own source)

Residue	Case studies	Electricity supplier	Current application	Stakeholders (beneficiaries)	Financing
Coffee husks (Minas Gerais)	1	Centralized (CEMIG[15])	Heat generation Chicken bedding	Coffee enterprise Chicken farms	Private
Rice husks (Minas Gerais)	1	Centralized (CEMIG)	Chicken bedding	Chicken farms	Private
Sugar cane bagasse (Minas Gerais)	1	Centralized (CEMIG)	Heat generation	Community	Private
Sawdust (Minas Gerais)	1	Centralized (CEMIG)	Chicken bedding	Chicken farms	Private

[15] CEMIG (Companhia Energética de Minas Gerais): electricity concessionary in the State of Minas Gerais

Sawdust (Pará)	2	Isolated	Heat and power Briquetting	Wood enterprise Community	CNPq[16] ELETROBRÁS

Zerriffi (2007) also analysed the long-term impacts of biomass schemes on the heat and electricity supply situation based on institutional and management variables:

- Location
- Quality of electricity service
- Sustainability
- Replicability

Sustainability has a broad meaning, which encloses environmental, socio-economic and technical dimensions. When applied to energy generation, this means that the combustion of biomass residues should be economically feasible, create social well-being and the environment should be kept in balance. The economical sustainability is represented by the regularization of the external and internal investments (Paz et al., 2007). Furthermore, it should meet the needs of the present generation without compromising the ability of future generations to meet their own needs (UN 1987). Table 6.2 lists the long-term impacts of selected biomass utilization schemes in Brazil.

Table 6.2: Long-term impacts of five biomass schemes in Brazil (own source)

Residue	Location	Quality of electricity service	Sustainability	Replicability
Coffee husks (Minas Gerais)	City	High	High (fertilization) Low (heat generation)	High
Rice husks (Minas Gerais)	Countryside	Low	High	High
Sugar cane bagasse (Minas Gerais)	Countryside	Low	Medium	High
Sawdust (Minas Gerais)	City	High	High	High
Sawdust (Pará)	Countryside	Low	High	High

[16] CNPq (Conselho Nacional de Desenvolvimento Científico e Tecnológico): research and technology council

Replicability is a measure of whether the characteristics of the particular biomass scheme can be used to provide heat and/or electricity to new costumers (Zerriffi, 2007). Sustainability in this sense means the ability of the biomass energy generation scheme to cover its costs and provide functioning systems over a long period of time.

6.1.1 Biomass utilization schemes in Minas Gerais

Minas Gerais is an agricultural State, with a long tradition in coffee and sugar cane cultivation in large-scale to small-scale farms. Rice is only cultivated for subsistence purposes, in small-scale. The carpentry and furniture industry is relatively strong in the region, since eucalyptus trees are broadly cultivated. The wood processing establishments are mostly located in the city of Vicosa, the commercial center of the adjacent rural areas and where the Federal University is located.

The investigated cases are small-scale farming and carpentry structures located in the region called Zona da Mata in Minas Gerais. There, 100 % of the urban population and most of the rural population have access to electricity. In all investigated cases, rural communities have infrastructure, health and education. The level of instruction of adults is up to the primary school. Nevertheless it was observed a change in the prevalent education patterns. Youngsters are having more schooling opportunities (high school) than their parents during the past. The average family composition is three children per household.

As Table 6.1 describes, coffee husk is a residue used for drying the coffee beans (heat generation). The sustainability of this biomass utilization scheme is low due to the low efficiency of furnaces for firing the husks and the bulkiness of the material, which hinders storage and transportation. However, the utilization of coffee husks for chicken bedding and further fertilization is currently employed in larger-scale with success. Similarly, rice husks and sawdust from Minas Gerais are also employed for chicken bedding, but not for heating purposes. Hence, the sustainability of the coffee husks, rice husks and sawdust for fertilization purposes is considered of high scale because of the recycling of nutrients in the soil and establishment of a market chain for residues through chicken farms.

The author could identify a high replicability of coffee, rice and sawdust schemes (Table 6.2). Several coffee and rice producers integrate coffee cultivation with chicken farming in the rural regions of Vicosa, aiming saving expenditures with chemical fertilization. It is important to note that the potential for using coffee husks, rice husks and sawdust for compaction and energy generation in Minas Gerais exists where the demand for chicken bedding is low and hence the marketing of husks and sawdust does not take place (personal communication).

On the other hand, sugar cane bagasse appears a promising biomass source for supplying the local demand of heat for organic sugar production in small-scale farms of Minas Gerais. The Organic Sugar Cooperative located in a rural community (Rio Branco) of the city of Vicosa cultivates sugar cane and produce organic sugar using bagasse for generating heat, which is used on the processing of sugar cane juice. The cooperative was founded in 2004 and counts with 10 sugar mills and 25 members. The rural community has approximately 200 inhabitants. There, a primary school, an event saloon and a small health centre are available.

The cooperative could not inform the author about favourable policies or subsides through government grants for using alternative fuels. Nevertheless, there is a perspective of external subsidizing. The governmental agency for agricultural development (EMATER) contacted the bank Santander, which offered a donation of 50,000 BRL for the farmer's cooperative. Each sugar mill could use 4,500 BRL strictly for improving the infrastructure of the mills. As reported, farmers are sceptical regarding the donation. They are afraid to be forced to pay back the amount of money, which was offered by the bank.

The average yield is 70 tones sugar cane per hectar, depending on weather conditions. Each ton of sugar cane originates 300 kg of bagasse. Each association member cultivates sugar cane in own farms, varying from 0.5 – 4 hectares of sugar cane per hectare. In total, the association cultivates 10 hectares, so it produces approximately 700 tonnes of sugar cane per year. From this amount, 200 tonnes of bagasse is obtained and used for cooking the sugar cane juice and producing brown sugar. The organic sugar is freighted with trucks to Belo Horizonte, the capital of Minas Gerais, and sold in rural markets. The transport costs 100 BRL / ton of sugar.

The mills operate using electrical energy, approximately 1.5 kWh per day. According to Pedro, the cooperative's leader, electricity arrived in his community only in 1987. However, there are still days when electricity fails. In Brazil, it is common that the quality of electricity service is much lower in rural areas than in urban areas (Table 6.2). This is due to the large extension of the grids for energy distribution.

Table 6.3 Decentralized biomass generation scheme using bagasse in Minas Gerais (own source)

Community Rio Branco	
Inhabitants	200
Cooperative members	25
Sugar mills	10
Cultivation	10 hectares
Production sugar cane	700 tons / year
Production bagasse	200 ton / year

The author visited the cooperative in 2008 and 2011. With interviews, a detailed description of the situation could reveal if changes in the quality of life have occurred. During the last visit, it was recognized that farmers succeed in constructing a building for the collection, storage and distribution of the organic sugar (Figure 6.1).

Figure 6.1: Formal sugar storage (left) and new storage and administration building (right)

Furthermore, they improved the installations for processing the sugar cane juice and transforming into sugar. In order to cook 500 liters of juice per day, it is needed to burn

200 kg of bagasse. A "home technology" developed by Pedro, allows the pre-heating of sugar cane juice before cooking (Figure 6.2). The "double pan system" can save a cooking time of two and a half hours.

Figure 6.2: Formal pan for juice cooking (left) and new "double pan" (right).

The cooperative cultivate sugar cane organically and also it could be expected to continue the commercialisation of organic sugar. This represents a good perspective for covering production costs and further functioning of the cooperative. However, the utilization of the bagasse for heating purposes is still inefficient. The storage of bagasse requires large rooms and farm's environment suffer from visual and material pollution. Since the heat generation from non-compacted bagasse could be considered inefficient, the sugar cane bagasse scheme presents a medium sustainability parameter regarding heating purposes (Table 6.2).

In all investigated biomass schemes of Minas Gerais there was a lack of commitment for producing electricity using biomass in combustion process. According to academics of the Federal University of Vicosa, farmers prefer to carry out the production processes in their farms using and paying for the electricity provided by the regional concessionary CEMIG.

The government program Luz para Todos (Chapter 2) implemented rural electrification in almost 100 % of the remote communities of Minas Gerais. Light and power is crucial for establishing production processes in the agriculture and industry. The access to electricity increases economical possibilities for the rural population, and consumers are ready to pay for it, rather than producing power themselves in small-scale biomass firing facilities.

The life quality and economic perspectives of rural population in Brazil can differ enormously regarding electricity availability. In regions where the grids do not arrive and electricity is not available, the population is conducted to use alternative solutions for generating power. In the Amazon region, local diesel generators provide electricity few hours a day and can only cover basic needs. Nevertheless, governmental initiatives through university, research centers and experts have been implemented for the utilization of local biomass in Amazon.

6.1.2 Biomass utilization schemes in Pará

The North region has complex ethnic, socio-economic and cultural conflicts. Most part of the remaining Brazilian indigenous population is located in this area and the conflicts generated by the intense attempts of land appropriation results in losses in their cultural patrimony. The electric sector has a history of attempts to occupy part of those lands for generation purposes, but they were not always successful (Paz et al., 2007).

It is estimated that in the Amazon region 80% of the unattended communities have less than 30 households (Amazonas Energia, 2009), and very few settlements have more than 100 inhabitants. Because of the distance from the main transmission and distribution lines, grid extension for these communities is economically unviable. Hence, there are still many communities in the Amazon that cannot be assisted by the conventional system.

The use of expensive diesel generators in this region is still the norm (OECD/IEA, 2010). The cost of electricity generation in these areas is high and reaches USD 200 / MWh (ESMAP, 2005). This is mainly due to the high costs of fuel and transportation and often because of the inefficient functioning of old motors. (Goldemberg et al., 2004. A liter of diesel transported to generate electricity may mean using 2 liters for obtaining it.

It was reported that the diesel electricity provided by CELPA is mainly used for entertainment, illumination, water pumping, food refrigeration and air-conditioning. However, it has been not possible to implement commercial activities like fruit industrialization, conditioning, freezing of agricultural and fishery products. Refrigeration for the local agriculture and fishery would improve community's ability to commercialize its products. However, these cannot be reach using diesel generators.

The investigated cases are two isolated communities located in the Marajó Island: community of Santo Antônio and community of Porto Alegre. In both communities, most of the rural population have access to electricity only few hours per day. Furthermore, there is a lack of basic infrastructure, like waste water treatment. Sewage canalization is directed to the rivers. Health service is only available in Breves. Education is secured only up to the primary school. Adults are mostly illiterate. Nevertheless it was observed a change in the prevalent education patterns. Youngsters are having more schooling opportunities (high school) than their parents during the past. The average family composition is ten to twelve children per household.

The local economy has a strong extractivist component, which is a cultural heritage from indigenous people and colonizers. The main economic activity is wood exploitation, with a low rate of industrialization. Secondary activities are cattle breeding, soy cultivation, boot transportation services, fishery and tourism. The carpentry and furniture industry is relatively strong in the investigated region. The wood processing establishments are mostly located in the city of Breves, the commercial center of the adjacent rural areas.

In the communities of Santo Antônio and Porto Alegre, the author investigated two biomass projects, which implemented small-scale biomass plants using direct combustion of wood residues in a CHP plant (Combined Heat and Power plant) using ORC (Organic Rankyne Cycle). The projects were financed by the Ministry of Education (CNPq) and ELETROBRÁS and executed by the Federal University of Pará (UFPA). The aim was to generate heat and power using wood residues from local sawmills.

The two isolated communities are located in the Marajó Island, where life is deeply connected with the numerous estuaries of the Amazon River (Figure 6.3 and Figure 6.4). The area has a wetland type of ecosystem. The main economic activities are fishery, wood milling, fruit production, rice cultivation, ice production and oil extraction from native seeds. The Amazon region has a high diversity of native oil seeds, with pharmaceutical and cosmetic importance.

Figure 6.3: Boat transportation (left). Ice and fish handling (right)

Figure 6.4: Typical house located in the Marajó Island (left) and daily life (right)

In 2009, the UFPA implemented the **Project ENERMAD** in the community of Porto Alegre. The CHP plant of 200 kW is fed with wood residues from milling (Figure 6.5). The lower heating value of the native wood is approximately 18 MJ / kg, the moisture is 40% and the load is 150 kg of sawdust per hour. The biomass residues produce steam, which is used to rotate the turbine and generate electricity. Attached to the power facility is a seed oil factory (Figure 6.6), which uses the waste heat produced from the saw wood residues to the oil press machinery. The seeds are supplied by community members, which earn 1.00 BRL / kg of collected seeds. The variety of seed species, like buriti and muru-murú, and the amount of oil extracted for each different seed is depending on the seasonality of seed production.

Figure 6.5: Integrated biomass utilization scheme: sawmill (left) and sawdust CHP plant (right).

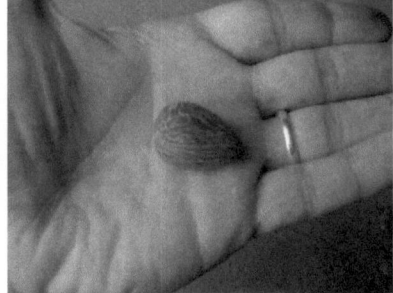

Figure 6.6: Integrated biomass utilization scheme: oil extraction (left) and muru-murú seed (right).

The project was implemented in four steps:

1. Civil engineering (store building)
2. Installation of equipments
3. Plant commissioning
4. Training of operators

The transport logistics and the lack of local infrastructure were the main difficulties faced during the execution of the project. The equipments were transported by ferryboat and delivered manually. The plant commissioning was performed by the project engineers together with the local operators, in each step of the production chain: combustion of sawdust, biomass drying, seed pressing and oil extraction. The training of operators involved: security in furnace operation, food manipulation, environment protection, business administration, operation and monitoring of the biomass CHP plant, ice factory and oil extraction facility. In addition, the project coordinators (UFPA) also offered a training course for the efficient use of electricity.

The Project ENERMAD has the support of a NGO for establishing a sustainable market for the seed oil through regional market managers. The inhabitants created a cooperative with the support of the university. The cooperative has 14 operators, which are distributed in the electricity generation (8 persons), oil extraction (4 persons) and ice factory (2 persons).

The electricity generated from sawdust is available for supplying households and commercial / industrial establishments. The electricity is measured in each house (electric meter) and charged monthly. The cooperative is responsible for the maintenance of the mini-grid, the electrical load assessment in each household and the delivery of electricity bills for the consumers.

The project coordinators (university) are responsible for the socio-economic monitoring of the activities, which occur every six months. The evaluation covers:

- Development of rice cultivation and productivity
- Native forest area and quantity of oil seed trees available
- Quantity of milled wood, quantity of sawdust generated
- Quantity of biomass consumed in the CHP plant
- Generated and consumed electricity
- Quantity of ice produced and sold

All these information contributes for a sustainable administration of biomass energy projects. The sustainability of the project (Table 6.2) is guaranteed by several factors:

- Availability of wood and wood residues
- Availability of fresh water (CHP plant, ice factory and further activities)
- Availability of native seeds (oil extraction)
- Costs of electricity (for consumers)

Approximately, 750 kg of biomass is being consumed in the CHP plant. Mostly, the supply is guaranteed by the local sawmill and partially by adjacent sawmills, which provides biomass at zero costs, and as well by the residues originated from the seed oil extraction. This availability ensures the full functioning of the plant. The costs of the biomass electricity is 140 BRL / MWh, which is very attractive, compared to the diesel electricity and also to the prices charged by the local concessionary CELPA[17], which is currently 360 BRL / MWh (ANEEL, 2011).

The integrated biomass plant is located in the riverside. Hence, fresh water is obtained by direct pumping. Afterwards, the water is purified in a treatment plant, which is integrated to the ice factory. The ice production can supply the community and the surplus is sold to adjacent localities. Before the project implementation, the fisherman used to travel around 120 km to obtain ice for conserving their fishes. Nowadays, they can buy local ice for a cheaper price and save money of transportation.

The local sawmill processes wood and sell the whole production in the local market, in the city of Breves. With the autarchic production of electricity, the sawmill tripled its daily production, from 5 m³ to 15 m³ of processed wood. Furthermore, the cooperative, supported by a NGO, is selling the extracted crude oil for approximately 10 BRL / liter to a cosmetic factory in Belém. This potential could be increased, since there is an immense variety of vegetable species. Furthermore, the processing and transesterification of crude vegetable oil into biodiesel requires further research.

Due to the availability of accessible electricity, there is a great perspective for initiating further local economical activities such as: fruit industrialization (açaí fruit), rice processing (peeling) and furniture factory. Hence, the focus on power generation combined with productive activities, like seed oil extraction, increases project's sustainability and also makes it less vulnerable to competition from the diesel generators, which are anyway focused on serving households.

Social improvements were also identified in the project. Sawdust electricity provides the community with light and power more hours per day. Hence, adults can participate in basic schooling programs offered by the government in the evening. Furthermore, social evenings are happening more often in the community.

The success of the Project ENERMAD was used to illustrate how small-scale decentralized biomass schemes could work. However, the isolation of communities and the lack of a minimum level of schooling and instruction of the locals could increase the

---

[17] CELPA: Centrais Elétricas do Pará S.A.

difficulties related to project management, operation and maintenance. As reported by Prof. Goncalo Rendeiro, the **Project Marajó**, in the community of Santo Antônio, failed due to business administration problems. This project was inaugurated in 2007 and the variables were similar to the Project ENERMAD (Table 6.4).

Table 6.4: Biomass projects implemented by UFPA in two communities (own source)

	Communities	
Project variables	Porto Alegre Project ENERMAD	Santo Antônio Project Marajó
Inhabitants	240	72
Diesel consumption	450 l / week	200 l / week
Wood production	8,500 ton / year	3,600 ton / year
Sawdust production	4,500 ton / year	1,800 ton / year
Sawdust consumption	750 kg / hour	750 kg / hour
Biomass plant capacity	200 kW	200 kW
Oil extraction capacity	100 kg of seeds / hour	100 kg of seeds / hour
Seed price	1 BRL / kg	1 BRL / kg
Oil price	10 BRL / liter	10 BRL / liter
Ice productivity	10 ton / day	10 ton / day
Electricity price (diesel)	1,30 BRL / kWh	1,30 BRL / kWh
Electricity price (biomass)	0,15 BRL / kWh	0,14 BRL / kWh
Household expenses with electricity (BRL and kWh / month)	BRL 27.00 / month 180 kWh / month	BRL 21.00 / month 150 kWh / month
Project workers	8	14

The integration of production processes was also the objective of the Project Marajó. A local sawmill should supply the plant with wood residues and generate heat for drying biomass and extracting vegetal oil from native seeds. The electricity generated from the steam should supply inhabitants, an ice factory and a commercial deep-freeze store. Since 2009 the project is not in operation due to difficulties in business and administration practice of the semi-analphabetic population and also due to corruption problems.

In both Amazonian case studies, the rural populations face harsh living conditions: long distances, difficulties in transportation and lack of money favour the use of local resources. Nevertheless, the author could identify a high replicability of biomass projects using sawdust (Table 6.2). Several sawmills in the Marajó Island see this initiative as a chance for improving life quality, production processes and income in the remote areas of Amazon.

6.2 Practical application of results into further projects

Significant features of energy systems are not only their condition of renewability, but also their environmental sustainability and social impacts. Generally, the diffusion of alternative technologies could have a positive impact on the welfare of individuals, economic development and social change. New commercial possibilities for agricultural biomass in the energy sector could stimulate farm economies, and revitalize struggling rural communities. However, the üpresent case studies have shown complexities and socio-economic and environmental implications of resource use for energy production. As observed, the implementation of decentralized biomass energy projects could be hindered due to:

- Broad distances leading to difficulties and high costs of transportation
- Absence of a biomass supply chain, complicating the diffusion of goods and services between communities
- Lack of basic infrastructure (sanitation, education and health)
- Low income and low energy consumption lead to small returns on investments
- Environmental degradation

Most of crops in South and Southeast are located at places where freight is expensive up to marine ports. In addition, Brazil's transportation system is relying on highways and roads, which in most regions of the country are in precarious situation. Hence, the high cost of transporting biomass by trucks hinders the commercialisation of biofuels.

It is known that some attempts of pellets production in Southeast Brazil for exporting failed due to high transportation costs. The author could obtain information about these efforts through interviews. The Fairbiomass Pellet Export project, located in Itajubá, Minas Gerais, planned to use coffee husks delivered from Ipanema Coffee, a large-scale coffee producer in Brazil, to produce pellets and export initially to the power plant Essent in Holland. According to Ricardo Pin, director from Fairbiomass Pellet Export, the constraints regarding the exportation of coffee husk pellets to Holland were due to Brazilian harbor bureaucracy and to problems on the certification of the coffee husks, which should be sustainably produced for exportation.

Brazilian ports are not equipped for fast carrying bulk biomass such as charcoal, bagasse, wood pellets or chips. Most of the ports are public and their services are expensive and inneficient. In order to solve this problem, private ports located at strategic places and conveniently equipped with belt-carriers have been used for wood chips exports. A least three big companies for both energy and pulp purposes have well structured and specialized ports such as Amcel, in Santana (State of Amapá) and Bianchini (Tanac), in state of Rio Grande do Sul (Walter et al., 2009).

The important aspects in relation to isolated areas are the small electricity demand, the lack of skilled people, and difficulties in properly operating and maintaining power equipment (Goldemberg et. al, 2003). Hence, one of the barriers for biomass use in isolated villages could be a matter of adapting technologies. As Rendeiro (2010) reported, the main factors that affect the economical success in sawmills are: wood conditioning, type of equipment used, wood species and working skills. In consequence, power systems for these areas must be of small capacity and as simple as possible.

Brazil, a major food exporter, is expected to increase the exports of food rapidly during the coming decades (Smeets et al. 2008). Hence, the amount of agricultural residues could be expected to augment as well, increasing the amount of available feedstock for heat and power generation. During this research, the author identified that the implementation of decentralized biomass energy projects in Brazil is currently depending on:

6.2.1 Technology

Biomass technologies should be adapted according to geographical and socio-economic conditions in Brazilian regions and Brazilian enterprises. Hence, the implementation of innovative projects using agricultural and forestry residues is depending on scientific research programs that targets the total use and recycling of biomass.

Recent studies (Junginger et al., 2008) showed that a series of new configurations and improvements in logistics would able to export pellets from Brazil with competitive international prices. As an example, fluvial transport may be a solution for the Northern and Center-Western region due the existence of several rivers appropriated for this purpose (e.g. Amazonas and Madeira river). For the other Brazilian regions, fluvial transport should be mixed with railway.

Biomass compaction through pelleting or briquetting might improve the transportation and storage of large amounts of biomass from a technical and economic perspective. Nevertheless, one significant challenge regarding the use of agricultural residues for energy generation is the technological improvement for small-scale combustion devices.

Besides the technological challenges, in Brazil, there are currently additional areas of concern that should be taken into consideration when stimulating new solid biofuels' markets. Some of the aspects to be taken into account are the lack of biomass solid fuel standards and the need of assessing the cost benefit of compacted biomass energy systems. Improvements in the logistics of the already established biomass networks are expected to contribute to costs reductions.

6.2.2 Socio-economic conditions

In order to safeguard small-scale agriculture and its agricultural production model, it has been suggested that biofuel production could be stimulated in a decentralized manner with scales allowing the entry of small-scale farmers as biofuel producers and the eventual reduction of dislodgment between production areas and consumer centres (BNDES / CGEE et al. 2008). Concerning economic concentration and its implications, a bioenergy system using biomass residues could be characterized as a highly decentralized industry compared with other energy related activities since an important number of suppliers would be involved.

In regions where the population doesn't have access to electricity, the alternative option is to use residues from agricultural activities, forest residues (branches, leaves, etc) and sawmill residues (sawdust, wood chips, etc) to generate heat and electricity. Local biomass projects should be financially sustainable and replicable, meeting the needs of locals. The organization of biomass production and supply chains involve stakeholders from all stages of the life cycle of coffee, rice, wood or sugar cane production. Another important issue is the choice of the best available energy generation scheme for the considered region.

There are several social benefits of decentralized biomass fired cogeneration. Regarding the large-scale plants, it has the potential of increase employment for populations located in the surroundings of the biomass facilities, for instance ethanol plants. On the other hand, it could broaden the scope and increase income for small-scale sugar cane farmers, which produce mostly organic sugar.

As sugar mills, wood mills also tend to be located in rural areas, near plantations. Hence, bagasse and wood cogeneration could provide a more efficient and regular access to electricity supply in areas distant from the grids. This facilitates the collection of electricity payments by electricity boards in rural areas. Simultaneously, an increase on the reliability of distributed power could enhance the life quality of rural populations (WADE, 2004).

The project aimed at serving the needs of isolated communities should be specifically drawn up for this population in accordance to their particular needs and potential. As described by the author, the project Buriti is focused on providing electricity to productive activities in order to improve economic output and social development. Other projects, like Santo Antônio, are more centered on household supply, allowing higher levels of electricity consumption compared to diesel generators.

The sawdust CHP plant integrated with the oil extraction facility fosters the substitution of fossil fuels by renewable sources that are available locally. Such substitution generates positives impacts on the local and national economy and reduces the emissions of GHG into the atmosphere.

6.2.3 Environment

Currently in Brazil, there is an ongoing debate about the land use exploitation and the effects of intensive crop cultivation, specially regarding the frontiers of sugar cane cultivation for ethanol production. In principle, the compaction of biomass residues in

Brazil would result neither in drastic land use changes nor in the already practiced agricultural activities. Therefore, it could be assumed that the feedstock required, as the straightforward inputs to the system are already available without causing processes disruption. Nevertheless, it is important to consider the whole supply chain of solid biofuels, in order to assess the socio-economic and environmental impacts of their production in all the stages involved. Hence, despite the production of residues being a natural process in agriculture and forestry systems, a life cycle assessment should be carried out.

Biomass fuels have several environmental advantages over fossil fuels, being a core one that it is a renewable resource, offering a sustainable and reliable supply. The energy sector has a small share in the total Brazilian emissions due to the relevant participation of renewable energy in the Brazilian energy matrix, mainly hydroelectricity and ethanol consumption in the transport sector.

There is a utopia of preserving Amazon's forests, hindering the socio-economic development of local populations. As a result, however, the region has been neither preserved nor developed, and suffers from destructive exploitation (Andrade et al. 2011). On the other hand, development could be compatible with environmental protection, when natural resources are used respecting the time necessary for recovering, considering the local culture and the needs and aspirations of the individuals.

Environmental Impact Assessment (EIA) is required only for medium and large projects for forestry or agriculture monoculture in Brazil. Most of small-scale projects do not carry out EIA. Thus, illegal exploitation of native forests and pollution of ecosystems is often recognized (Walter et al., 2009). Nevertheless, a legal framework was enforced in 1998 and set sanctions for environmental crimes.

Biomass fuels have lower emissions of particulates, SO_2, NO_2 and CO_2 compared to coal and other fossil fuels. Biomass combustion emits less GHG than composting. Besides, when decomposing, biomass can release methane, a GHG 27 times more potent than CO_2 (WADE, 2004). Furthermore, the cogeneration process using biomass is more efficient compared to conventional combustion process, which do not recover heat.

In Brazil, there is much evidence that areas covered with native vegetation are cleared out to grow soy, sugar cane, rice, and eucalyptus wood. Hence, bioenergy obtained though large-scale monoculture systems could be rather a threat as an opportunity for ecosystems and people (FNR, 2009). On the other hand, farmers could cultivate energy crops in small-scale systems and diversify the species varieties. The desired amount of biomass used for energy generation could be reached through cooperation schemes.

If the forests that provide wood fuel are re-replanted at the same rate as they are cut down, then such fuel use should in principle be sustainable. Bioenergy systems are considered sustainable when they maintain their production over the long term without overt depletion of the resources that originally gave rise to them, such as biodiversity, soil fertility, and water resources. Such focus is based on one of the classical definitions of sustainability: "the amount of production that can be sustained indefinitely without degrading capital stocks, including natural capital stocks" (UN, 1987).

When forests are managed sustainably, the CO_2 absorbed in growing replacement trees should equal the CO_2 given off when the original trees are burned. However, this is only true when complete combustion of the wood occurs and all the carbon in the wood is released as carbon dioxide (Boyle et al., 2003).

The major challenge in many countries is to run a proposed bioenergy project through contractual and consenting processes that are needed to reach commercial reality. Even when biomass conversion is considered a state-of-the-art technology, the socio-economic potential is no longer fully realized in many countries of the world. Hence, there are several issues to be resolved:

- Land availability
- Environmental impacts of monoculture in indigenous forests
- Sustainable cultivation of energy crops and plantation forests
- Storage and transport of large volumes of biomass residues
- Energy output/input ratios
- Economic competition from cheap fossil fuels

In Brazil, deforestation is unlikely to decline in the near future despite low population density in the Amazon and Cerrado regions. High food and fuel prices will favour continued forest clearance for production of livestock and agricultural crops for food, feed and biofuel to meet the global demand. Sustainable forest management will continue to be a challenge in a number of countries where land tenure is poorly defined (FAO, 2009). Federal policies should attempt to ensure the sustainability of woody biomass harvesting. This should be done coupled with measures towards winning the public acceptance.

6.2.4 Policies

The Brazilian policy framework supports the development of renewable energy markets in the electricity sector. Therefore, the electrification of rural areas with energy from off-grid systems based on domestic resources; could contribute to achieve environmental and social targets such as the inclusion of rural communities in the development of the agro-energy sector in the country. In addition, the participation of small holders as market actors could boost the social and economic benefits in rural and remote areas.

Policies for renewable energy and the administration of agricultural and manufacturing activities in Brazil are expected to play a crucial role regarding their contribution to sustainable development. The stability of economic incentives for renewable electricity sources; the creation of institutional structures to assure the environmental performance of decentralized alternatives. Besides the market barriers that have to be overcome, the social and environmental dimensions should be considered, in order to define the pathway to take advantage of the native resources in sustainable energy systems.

Worldwide, the main economic barriers to renewable energy projects include high initial costs and the small-scale production of equipment and systems. To overcome these barriers the creation of a market of minimum size is essential. The successful implementation of renewables has been based on tax incentives, but the Brazilian government has never formulated a comprehensive and long-term policy for renewables with this kind of incentive. Instruments such as tax reduction for imported devices of higher efficiency, credits on taxable income and accelerated depreciation have also been helpful (Goldemberg et. al, 2003).

In general terms, regulatory actions by ANEEL address important issues, but there are many doubts concerning their effectiveness as tools for fostering renewable electricity in Brazil. In the event that the mandatory market is approved with no corresponding action regarding economic and science and technology policies, an external dependence on equipment suppliers will be created in several renewable energy sectors (Goldemberg et. al, 2003).

Ten years ago power generated from alternative plants could not be sold to the grid. Hence, only after the implementation of energy policies the commercialisation of electricity was possible. Nevertheless, decentralized energy production in Brazil faces institutional barriers. The policies and measures created by the government, such as PROINFA, are still unfavourable for the success of electrification projects using biomass in rural areas.

One of the main policies of PROINFA is implementing grid extension to remote areas. Connection costs may become higher if population density is low and consumers are dispersed (FAO, 2001). Therefore, decentralized renewable electrification options other than grid extensions, by either individual systems or mini-grids, may provide much cheaper alternatives particularly in the remote and isolated villages of the Amazon region (ESMAP, 2005).

6.2.5 Financing: R&D Projects

Lack of continuity on biomass projects, biofuel exports and long-term contracts repels investments. The establishment of a domestic market for solid biomass residues in Brazil could be an alternative for such barriers. As it happened with the ethanol program, large-scale production and business maturity can only be achieved through long run initiatives.

The Brazilian government has established that resources between 0.75 % and 1 % on net sales of generation concessionaries, transmission and distribution of electricity, would be used to fund programs and projects in the energy area. The idea was to use the expenditures of enterprises in R&D projects for implementing alternative energy sources with lower costs and better quality, waste reduction, energy efficiency, besides stimulating competitiveness of national industrial technology (Pereira et al., 2010).

According to FINEP (2011), the percentage of annual revenues dedicated to R&D projects varies from 0.5 % (distributing utilities) to 1 % (transmission and generating utilities). The half of these values is spend by the utilities themselves in internal R&D projects. The other half is collected by a governmental fund, so called CT-Energ. Distributing utilities have to invest their total annual revenues (0.5 %) in end-use energy efficiency projects.

The Brazilian Development Bank BNDES[18] is the main financial support instrument for investments in all economic sectors. The Bank allocates special resources, preferably in the form of long-term funding and shareholdings, in addition to supporting undertakings that contribute to economical and social development (BNDES, 2011). BNDES invest not only in large-scale projects, but also in undertakings by micro, small and medium-sized companies, individuals and public administration agencies. Requests can be made directly with the BNDES or with an agent institution. The majority of operations are carried out indirectly through a partnership with a network of accredited financial institutions located nationwide. These agents may be commercial, public or private banks, development agencies or cooperatives. In indirect operations, the Bank reallocates financial resources to accredited agents, which are responsible for credit analysis and approval.

The BNDES make available supporting mechanisms such as: financing and non-reimbursable resources. Financing schemes grant investment projects, isolated acquisition of new machinery and equipment, exports of machinery, Brazilian equipment and services and the acquisition of goods and production inputs. Additionally, the BNDES offers financing programs of a transitional nature, which are focused on a determined economic segment. The Bank also administers some funds, which are resources to support specific activity sectors. The non-reimbursable resources are investments of a social, cultural (educational and research), environmental, scientific or technological nature, which need not be financially reimbursed to the BNDES.

The financing of biomass energy generation projects could also be executed by international research cooperations. The German Ministry of Environmental, Nature Conservation and Nuclear Safety (BMU) and the KfW Development Bank just stared establish an innovative climate protection fund. This pool assists developing economies to invest in renewable energy and measures for energy use efficiency. With regard to

---

[18] BNDES: Banco Nacional do Desenvolvimento

the objectives of R&D projects for sustainable energy supply in Brazil this might be one of many possible funding sources.

Another option could be the Investment Bank Berlin (IBB), which is running a funding program to strengthen small and medium sized companies. In focus of the program are evolvement activities to enter foreign markets. Here is a potential pool for German project partners. The German Ministry of Economics and Technology runs the initiative Germany Trade & Invest (GTAI), which belongs to the program "Technology for climate protection and energy use efficiency". Here technology co-operation of small and medium sized companies with foreign countries are supported. This might give another option for German partners to increase funding.

Furthermore, the BMBF is running the program "Research in sustainable development". Here the branch energy use efficiency, renewable energy, based on research and development will be examined as potential source. Research institutions and companies can be supported.

6.2.6 Participation of stakeholders

Since biomass residues are transported in Brazil commonly in their natural bulky state, the establishment of a Brazilian network for biomass compaction could provide costs reductions in the upgrading process of residues (UNICA, 2010). Another influential factor for the development of for further compacted biomass market opportunities -both in heating and in electricity production- is the price.

Technical assistance and training should be provided on a long-term basis. National and international donors, such as universities, NGOs, government and energy agencies, must implement energy generation projects towards creating access to resources and supporting efforts to create markets and promoting community's autonomy (Zerriffi, 2007). Furthermore, the strategy for implementation of networks should take into account the characteristics of the specific community and the benefits from the utilization of local resources, in this case biomass residues, should be shared equally among the local population.

The participation of rural communities is a fundamental key to reaching socio-economic and ecological transformation. According to GTZ (1999), *"participation is an interactive and cooperative process of analysing, planning and decision-making. It allows all participants to formulate their interests and objectives in a dialogue, which leads to decisions and activities in harmony with each other, whereby the aims and interests of other participating groups are taken into account as far as possible"*. The author noted that communication skills aiming the involvement and participation of stakeholders in rural areas are fundamental in communities where the level of illiteracy is high.

However, there is considerable resistance to organised participation in Brazilian rural areas. Farmers are in most cases sceptical about new ideas. Since farmers devote all of their energy to the daily life for survival, they can only approach to long-term activities if this is achieved.

6.3 Discussion

Changing established social and economic systems in Brazilian rural areas is extremely complex. This is partially due to the geographical isolation and to the low level of illiteracy. In some cases, the lack of opportunities offered by the government, like infrastructure, education and health, hinders the organisation of new socio-economic arrangements.

The author could identify that the investigated sugar cane cooperative in Minas Gerais improved facilities and installations aiming energy efficiency. According to Pereira et al. (2010), with an increase of income, families tend to opt for more modern methods of cooking, heating, lighting and use of home appliance products. The discussion with the cooperative's members showed a great interest in implementing new technologies for bagasse utilization and ethanol production. It was reported that there exists a strong cooperation between the University of Viçosa and the farmers aiming ethanol production for generating heat and electricity.

The local populations, who have limited access to information and education, have less participation in the definition of priorities for the societies in which they live (Andrade et al., 2011). Therefore, public intervention such as from universities and research institutes is important for supporting stakeholders in rural areas. However, this depends on financial support and as well on the preferences of the potential consumers.

The presence of subsides and governmental financial support influences the viability of a decentralized electrification scheme. It has positive and negative effects. It could improve the finance of the biomass scheme or it could make it less competitive with the current electricity scheme available. Since low-income costumers are allowed paying lower electricity bills, there is no willingness to invest in alternative production of electricity from local resources in rural areas. As an example, in Minas Gerais, it is safer and cheaper to use the available centralized energy service supplied by the hydropower plants.

In Amazonas, the diesel subsidy enables the supply of electricity to isolated communities. On another hand, it inhibits the competitiveness of renewable sources of electricity in the region, as it alters the relative price of the options. The implication of eliminating completely the subsidies for diesel consumption and low-income consumers is that rural residents may be not able to pay the afforded electricity price and be excluded from the grid (Zerriffi, 2007). One way of modifying the current scenario would be to use the diesel subsides to enable the viable supply of renewable energy to isolated communities, on a long term and sustainable basis, rather than subsidizing fossil fuels (Andrade et al., 2011).

Greater chances for participation of stakeholders in local energy generation should be given. Communities could participate fully without worrying about operating illegally or being undercut by the large utilities. They could opt for electrification options that are best suited to their needs without loosing the low-income subsidies. Finally, governments could meet their desired electrification targets while supporting rural electrification goals (Zerriffi, 2007).

6.4 References

Amazonas Energia (2009) Available at: http://www.amazonasenergia.gov.br/cms/ (Last acess on June 15, 2011)

Andrade CA., Rosa LP. and Silva NF. (2011) Generation of electric energy in isolated rural communities in the Amazon region. A proposal for the autonomy and sustainability of the local populations. In: Renewable and Sustainable Energy Reviews 15: 493-503. Elsevier

ANEEL - Agência Nacional de Energia Elétrica (2011) Tarifas residenciais. Available at: http://www.aneel.gov.br/area.cfm?idArea=493&idPerfil=4 (Last acessed on Mai 27, 2011)

Azanha M. (2007) O mercado de trabalho da agroindústria canavieira: desafios e oportunidades (in Portuguese). Departamento de Economia, Administração e Sociologia – ESALQ/USP – Piracicaba, SP. Econ. aplic., São paulo, v. 11, n. 4, p. 605-619.

Barros EV. (1994) Princípios de ciências sociais para a extensão rural. Vicosa, UFV. 715 p.

Bartocci L. (2009) Perfil da mão de obra no setor sucroalcooleiro: Tendências e perspectivas (in Portuguese). Tese de Doutorado: Programa de Pós-Graduação em Administração, Departamento de Administração. Faculdade de Economia, Administração e Contabilidade, Ribeirão Preto – USP

BNDES – Banco Nacional do Desenvolvimento (2011) Available at: http://www.bndes.gov.br/SiteBNDES/bndes/bndes_en/ (Last accessed on 11 June 2011)

BNDES, CGEE, CEPAL, FAO (2008) Sugarcane based bioethanol: Energy for sustainable development. Available at: http://www.bioetanoldecana.org/ (Last access 8 June 2011)

Boyle G., Everett B. and Ramage J. (2003) Energy systems and sustainability: Power for a sustainable future. Oxford University Press Inc, New York

Cardoso T. (2010) Cenários tecnológicos e demanda da capacitação da mão-de-obra do setor agrícola sucroalcooleiro paulista (in Portuguese). Dissertação de Mestrado: Universidade Estadual de Campinas (UNICAMP), Faculdade de Engenharia Agrícola.

ESMAP (2005) Brazil background study for a national electrification strategy – aiming for universal access. World Bank, Washington D.C.

FAO – Food and Agriculture Organization (2001) Environment and Natural Resource Working Paper. Available at: http://www.fao.org/DOCREP/003/X8054E/x8054e06.htm (Last accessed: 11 June 2011)

FAO - Food and Agriculture Organization of the United Nations (2009) State of World's Forests. Available at: ftp://ftp.fao.org/docrep/fao/011/i0350e/i0350e.pdf (Retrieved on July 9, 2010).

FINEP – Finaciadora de Estudos e Projetos (2011) CT-Energ: Fundo Setorial de Energia Available at: http://www.finep.gov.br/fundos_setoriais/ct_energ/ct_energ_ini.asp Last accessed: 11 June 2011

FNR e.V. - Fachagentur für Nachwachsende Rohstoffe (2009) Bioenergy http://www.fnr-server.de/ftp/pdf/literatur/pdf_330-bioenergy_2009.pdf

Goldemberg J. (2002) The Brazilian Energy Initiative - Support Report Presented at the World Summit for Sustainable Development, Johannesburg.

Goldemberg J., Rovere EL. and Coelho ST. (2003) Expanding access to electricity in Brazil. Available at: http://www.afrepren.org/project/gnesd/esdsi/brazil.pdf (Last access 31 March 2011)

Gomes J. (2010) O canavial como realidade e metáfora: leitura estratégica de trabalho penoso e da dignidade no trabalho dos canavieiros de Cosmópolis (in Portuguese). Institute of Psychology, Social Psychology, USP.

GTZ – Geselschaft für Technische Zusammenarbeit (1999) Land use planning: methods, strategies and tools. Wiesbaden Universum Verlag, 223 p.

Hassuani J., Leal M. and Macedo I. (2005) Biomass power generation: sugar cane bagasse and trash. Série Caminhos para Sustentabilidade, PNUD-CTC, Piracicaba. ISBN 85-99371-01-0

Junginger M., Peksa M., Ranta T., Rosillo-Calle F., Ryckmans Y.,Wagener J., Walter A., Woods J. (2006) Opportunities and barriers for sustainable international bioenergy trade. Sustainable International Bioenergy. Task 40. Available at: http://igitur-archive.library.uu.nl/chem/2007-0628-202122/NWS-E-2006-235.pdf

OECD/IEA (2010) Comparative study on rural electrification policies in emerging economies. Available at: http://www.iea.org/papers/2010/rural_elect.pdf (Last access 31 March 2011)

Olfert K. (2008) Kostenrechnung. s.l. : Kiehl, 2008.

Paz LR., Silva NF. and Rosa LP. (2007) The paradigm of sustainability in the Brazilian energy sector. In: Renewable and Sustainable Energy Reviews 11: 1558-1570. Elsevier

Pereira MG., Freitas MA. and Silva NF. (2010) Rural electrification and energy poverty: empirical evidences from Brazil. In: Renewable and Sustainable Energy Reviews 14 (4): 1229-1240

Portal Brasil (2011) Available at: http://www.portalbrasil.net/salariominimo.htm (Last access 15 June 2011)

Rendeiro G. (2010) Geração de energia elétrica em localidades isoladas na amazônia utilizando biomassa como recurso energético. Tese de doutorado. Programa de Pós-Graduação em Engenharia de Recursos Naturais da Amazônia - PRODERNA. Instituto de Tecnologia.Universidade Federal do Pará. Agosto 2011.

Shikida P. (2010) The economics of ethanol production in Brazil: a path dependence approach. Available at: http://urpl.wisc.edu/people/marcouiller/publications/URPL%20Faculty%20Lecture/10Pery.pdf (Last access 15 June 2010)

Smeets E., Junginger M., Faaij A., Walter A., Dolzan P. and Turkenburg W. (2008) The sustainability of Brazilia, An ethanol - An assessment of the possibilities of certified production. Biomass and Bioenergy 32: 781-813 Elsevier

UNICA - União da Indústria de Cana-de-açúcar (2010) Setor sucroenergetico registra Available at: http://www.unica.com.br/noticias/show.asp?nwsCode={E52DBED8-BA81-4FDB-9EDA-37590AB18DDE} (Last Access 16 June 2010)

UN – United Nations (1987) Our Common Future, Chapter 2: Towards Sustainable Development http://www.un-documents.net/ocf-02.htm#I

WADE – World Alliance for Decentralized Energy (2004) Bagasse cogeneration: global review and potential. Available at:

http://cdm.unfccc.int/filestorage/2/K/J/2KJXDVUHFZ0MYG3WOBT91NA7QE6LP5/4057%20Annex%206%20Bagasse%20Cogeneration%20-%20Global%20Review%20and%20Potential.pdf?t=cmF8MTMwNTU0MjkyMi40OQ==|SYSVFPJYJyez_gsY1e70e2aM_bs= (Last access 16 April 2011)

Walter A., Dolzan P., Quilodrán O., Garcia J., da Silva C., Piacente F., Segerstedt A. (2008) A Sustainability Analysis of the Brazilian Bio-ethanol. Unicamp. Department for Environment, Food and Rural Affairs (DEFRA), British Embassy, Brasilia.

Walter A., Dolzan P. and Piacente E. (2009) Biomass Energy and Bio-energy Trade: Historic developments in Brazil and Current Opportunities. In: Country Report: Brazil – Task 40 – Sustainable Bio-energy Trade; securing Supply and Demand. Available at: http://www.bioenergytrade.org/downloads/brazilcountryreport.pdf (Last access 16 April 2011)

Zerriffi H. (2007) From acaí to access: distributed electrification in rural Brazil. In: International Journal of Energy Sector Management 2: 90-117

7 Conclusions

Brazil has an enormous potential to generate energy from biomass residues. However, there is a need for combining different available renewable resources in the country. For instance, hydropower and sugar cane bagasse cogeneration could be complementary energy sources. Further initiatives supporting the access to electricity through the utilization of local biomass could open the scope from centralized energy generation to decentralized generation.

The broadening of energy generation from agricultural and forestry residues face several barriers. The diversity of situations described in this thesis shows that regionalism and culture plays an important role in the implementation of solutions and alternative technologies. The unequal income and electricity distribution of Northern and Southern citizens leads to a distinct ways of perceiving the environment and achieving socio-economic development. On one hand, the provision of energy allows socio-economic development, on another hand; it can create environmental impacts in the original ecosystems.

Historically, the Brazilian energy policies privileged the attendance of the great industrial consumer's needs and the great cities of the Southeast region. In spite of the unquestionable success obtained by this policy in fomenting the industrialization and expanding Brazilian GDP, it also hindered socio-economic development of remote areas. Hence, small-scale energy generation using biomass is still a challenge for several rural communities in the whole country, which face the lack of electricity supply and job opportunities.

The challenge in Amazonia is how to use resources to promote socio-economic development without impacting the balance of the environment. Sawdust utilization for energy generation could be considered sustainable if sawmills are operating in legal conditions. The control of deforestation in Amazonia is still a challenge, which involves corruption and idealism.

The development of rural communities in Brazil has been by the geographical barriers of a vast country. Individuals are strongly connected to cultural patterns[19], which have been traditionally and orally preserved through many generations of social isolation. In isolated communities with distinct ethnical and cultural characteristics, the planning of energy generation alternatives should be done through long-term actions. These should consider the sustainable use of resources combined with shared benefits to the local population, such as health, education and employment.

Even when from the economical and technological point of view, biomass power generation is considered feasible, the implementation of decentralized electricity production in rural areas depends on social inclusion and on the commitment of the locals. Biomass power generation should be combined with advice, instruction and efficient use. Finally, sharing management and responsibilities in rural development projects may involve multiple links among communities, government agencies and non-profit organisations.

The technological option to be adopted in rural areas could succeed more efficiently when coupled with governmental and institutional support. Such financing should cover initial costs, maintenance of biomass facilities and workshops for promoting local participation and economic development. Successful examples of local biomass utilization in isolated areas of the Amazon forest were recorded in Marajó Island. As the author reported in Chapter 5, biomass projects implemented and sponsored by governmental institutions, like the Federal University of Pará and ELETROBRÁS, are in most of the cases economically feasible. However, the technical and economical functioning of the biomass facilities is depending on the commitment and the level of illiteracy of the local population.

With rising electricity prices and increasing demand for renewable energy, biomass-fired combined heat and power (CHP) systems become more attractive. It is crucial that biomass resources are used efficiently. While CHP itself represents a significant efficiency gain, there are a number of other steps that can be taken to improve the

---

[19] Culture is everything humankind adds to nature, in the process of adaptation to the physical environment and to the social environment (Barros, 1994).

efficiency of existing and new steam CHP systems. These include steam system and boiler measures, waste heat recovery, and biomass drying and dewatering.

There is a large potential for improving the efficiency of electricity generation by introducing new technologies. Optimising the running of such systems requires taking into account not just fuel costs, capital expenditure and the consequences of anti-pollution legislation, but also wider notions of national energy self-sufficiency and expectations of economic growth.

Brazil's transportation infrastructure and port facilities are deficient and costly. The average distance to ports from Brazil's Center-West is over 1,000 kilometers, and Brazil still relies predominantly on expensive overland truck transportation to move most bulk commodities. Compacted forms of biomass such as wood pellets and briquettes could reduce transportation costs, depending on the economic feasibility of the compaction facilities. According to the results, briquetting could be a more viable option for Brazil. It does require lower level of technological skills and it can be managed more simply than pellet devices.

In both ways, with self-financing and governmental support ("grand fathering"), the establishment of biomass utilization schemes is dependent on stimulating local production for supplying biomass chains and markets. Hence, it is crucial to guide and assist the rural population in finding out the relation among residue utilization, environmental protection and income generation. In Brazil, most of the Federal Universities and research institutes like EMBRAPA have pilot projects, which are extended to the rural areas. The aim is to bring technology innovations to the field and improve living conditions of rural communities.

Considering the variety of Brazilian regions and cultures, energy generation technologies should be adapted to the resources available and the local socio-economic conditions. Investments in technology can equip people with better tools and make them more productive and prosperous, raising incomes and building capacity for future innovations (Paz et al., 2007). This was recognized in a case study in the Federal State of Minas Gerais (Chapter 6), where farmers developed their own technological solutions, using available resources, to improve industrial processes and increase income.

The investigation of six specific case studies gave the author the possibility of identifying different resource management strategies, which biomass technology was applied and the needs of the local population. The results show the potential of utilization of local biomass for energy generation in Minas Gerais and Amazonas. It was noted that the cultural, geographical and socio-economic discrepancies of Brazil play an important role in the implementation of future projects.

Several projects have been succeeding in establishing local energy generation from biomass resources. In most of the case studies there is a need for implementation on a commercial scale. Therefore, each community should develop its own concept according to specific characteristics and potential. Nevertheless, introducing new ideas into rural communities requires a deep interaction with stakeholders. The process of interaction among community members and experts should influence the development of activities. To conclude, it could be added that energy supply and consumption should be addressed through integrated research. The cultural, geographical and socio-economic discrepancies of Brazil could involve challenges on the implementation of future projects, even when their development would provide benefits for Brazil as a whole.

8 Annexes

Annex 1: Pellet Norms (DIN, 2010)

Parameter/Normen	DIN 51 731 (DIN 2008)	Önorm M7135	DIN EN 14961-2 (DIN 2010)
Diameter	4 to 10 mm	4 to 10 mm	6 ± 1 mm
Length	<50 mm	<5xd	3.15 to 40 mm
Bulk density	>0.65 kg/dm3	>0.60 kg/dm3	>0.60 kg/dm3
Moisture content	<12%	<10%	<10%
Abrasion	No requirements	<2.3%	2.3% - 3.5%
Ash content	<1.5%	<0.5%	0.7% - 3.0%
Calorific value	17.5-19.5MJ/kg	>18 MJ/kg	16 - 19 MJ/kg
Sulphur content	<0.08%	<0.04%	<0.04%
Nitrogen content	<0.3%	<0.3%	<1.0%
Chlorine content	<0.03%	<0.02%	<0.03%
Arsenic	<0.8 mg/kg	No specified	<1 mg/kg
Lead	<10 mg/kg	No specified	<10 mg/kg
Cadmium	<0.5 mg/kg	No specified	<0.5 mg/kg
Chromium	<8 mg/kg	No specified	<10 mg/kg
Copper	<5 mg/kg	No specified	<10 mg/kg
Mercury	< 0.05 mg/kg	No specified	< 0.1 mg/kg
Zinc	< 100 mg/kg	No specified	< 100 mg/kg
Binding agent	Not allowed	<2%	<2%

Annex 2: Brazil - political map (IBGE, 2011)

Annex 3: Increase in sugarcane planted in area in Brazilian regions from 1998 to 2008 (IBGE, 2011)

Region	Area planted 1998 (ha)	Area planted 2008 (ha)	% Increase
North	14.965	28.016	87,21
Northeast	1.251.348	1.277.481	2,09
Centerwest	367.600	888.311	141,65
Southeast	3.059.432	5.367.621	75,45
South	356.608	649.448	82,12
Total	5.049.953	8.210.877	62,59

Annex 4: Rice, coffee and sugar cane production in Brazil 2010 and projection for 2011 (IBGE, 2011)

Rice	2010	2011	Variation (%)
Area harvested (ha)	2,705,730	2,774,608	2.5
Production (t)	11,325,672	12,859,611	13.5
Average yield (kg/ha)	4,186	4,635	10.7

Coffee	2010	2011	Variation (%)
Area harvested (ha)	2,159,544	2,147,518	-0.6
Production (t)	2,862,013	2,598,935	-9.2
Average yield (kg/ha)	1,325	1,210	-8.7

Sugar cane	2010	2011	Variation (%)
Area harvested (ha)	9,191,255	8,869,136	-3.5
Production (t)	729,559,596	705,823,063	-3.3
Average yield (kg/ha)	79,375	79,582	0.3

Annex 5: Electricity consumption in Brazil

Regions	Electricity consumption (GW h)			Participation in the consumption in 2008 (%)
	1999	2003	2008	
North	3,604	3,956	5,036	5.27
Northeast	11,948	11,859	16,515	17.28
Southeast	47,283	41,743	51,479	53.86
South	12,667	12,963	15,454	16.17
Center-West	5,828	5,623	7,100	7.43
Total	81,330	76,144	95,585	100

Annex 6: Electricity consumption by sector in Brazil (in percentage)

Sectors	Electricity consumption (GW h)		
	1999	2003	2008
Energy sector	3.3	3.5	4.3
Residential	25.7	22.3	22.3
Commercial	13.8	14.1	14.6
Public	8.9	8.7	8.1
Agricultural	4	4.2	4.3
Transportation	0.4	0.3	0.4
Industrial	43.9	47.0	46.1
Total	100	100	100
Final consumption (10^3 toe)	27,144	29,430	36,830

The Brazilian residential sector answered for 22.3% of the total electricity consumption being distributed in 2008 (MME 2009).

Annex 7: Electricity price in different Brazilian regions (ANEEL, 2011)

Abbreviation	Concessionary	Residential (R$/kWh)	Duration of validity
AES-SUL	AES SUL Distribuidora Gaúcha de Energia S/A.	0.31497	19/04/2011 until 18/04/2012
AmE	Amazonas Distribuidora de Energia S/A	0.30425	01/11/2010 until 31/10/2011
AMPLA	Ampla Energia e Serviços S/A	0.40188	15/03/2011 until 14/03/2012
BANDEIRANTE	Bandeirante Energia S/A.	0.32537	23/10/2010 until 22/10/2011
Boa Vista	Boa Vista Energia S/A	0.26876	01/11/2010 until 31/10/2011
CAIUÁ-D	Caiuá Distribuição de Energia S/A	0.29764	10/05/2011 until 09/05/2012
CEB-DIS	CEB Distribuição S/A	0.27952	26/08/2010 until 25/08/2011
CELESC-DIS	Celesc Distribuição S.A.	0.32499	07/08/2010 until 06/08/2011
CELG-D	Celg Distribuição S.A.	0.29353	12/08/2010 until 11/08/2011
CEMIG-D	CEMIG Distribuição S/A	0.38978	08/04/2011 until 07/04/2012
ELETROCAR	Centrais Elétricas de Carazinho S/A.	0.36920	29/06/2010 until 28/06/2011
CERON	Centrais Elétricas de Rondônia S/A.	0.35123	30/11/2010 until 29/11/2011
CELPA	Centrais Elétricas do Pará S/A. (Interligado)	0.36990	07/08/2010 until 06/08/2011
CEMAT	Centrais Elétricas Matogrossenses S/A. (Interligado)	0.41257	08/04/2011 until 07/04/2012
COCEL	Companhia Campolarguense de Energia	0.33214	24/06/2010 until 23/06/2011
ELETROACRE	Companhia de Eletricidade do Acre	0.37545	30/11/2010 until 29/11/2011
CEA	Companhia de Eletricidade do Amapá	0.19729	30/11/2010 until 29/11/2011
COELBA	Companhia de Eletricidade do Estado da Bahia	0.38203	22/04/2011 until 21/04/2012
CELTINS	Companhia de Energia Elétrica do Estado do Tocantins	0.41807	04/07/2010 until 03/07/2011
CEAL	Companhia Energética de Alagoas	0.33363	28/08/2010 until 27/08/2011
CELPE	Companhia Energética de Pernambuco	0.34427	29/04/2011 until 28/04/2012
CERR	Companhia Energética de Roraima	0.32728	01/11/2010 until 31/10/2011
COELCE	Companhia Energética do Ceará	0.40199	from 22/04/2011
CEMAR	Companhia Energética do Maranhão (Interligado)	0.41392	28/08/2010 until 27/08/2011
CEPISA	Companhia Energética do Piauí	0.37314	28/08/2010 until 27/08/2011
COSERN	Companhia Energética do Rio Grande do Norte	0.34472	22/04/2011 until 21/04/2012

CEEE-D	Companhia Estadual de Distribuição de Energia Elétrica	0.31642	25/10/2010 until 24/10/2011
CFLO	Companhia Força e Luz do Oeste	0.30410	29/06/2010 until 28/06/2011
CHESP	Companhia Hidroelétrica São Patrício	0.38426	12/09/2010 until 11/09/2011
CJE	Companhia Jaguari de Energia	0.30617	03/02/2011 until 02/02/2012
CLFM	Companhia Luz e Força Mococa	0.42706	03/02/2011 until 02/02/2012
CLFSC	Companhia Luz e Força Santa Cruz	0.39938	03/02/2011 until 02/02/2012
CNEE	Companhia Nacional de Energia Elétrica	0.32818	10/05/2011 until 09/05/2012
CPEE	Companhia Paulista de Energia Elétrica	0.40444	03/02/2011 until 02/02/2012
CPFL-Paulista	Companhia Paulista de Força e Luz	0.32883	08/04/2011 until 07/04/2012
CPFL- Piratininga	Companhia Piratininga de Força e Luz	0.31421	23/10/2010 until 22/10/2011
CSPE	Companhia Sul Paulista de Energia	0.38596	03/02/2011 until 02/02/2012
SULGIPE	Companhia Sul Sergipana de Eletricidade	0.38054	14/12/2010 until 13/12/2011
COOPERALIANÇA	Cooperativa Aliança	0.35786	14/08/2010 until 13/08/2011
COPEL-DIS	Copel Distribuição S/A	0.30000	24/06/2010 until 23/06/2011
DMEPC	Dep. Municipal de Eletricidade de Poços de Caldas	0.30642	28/06/2010 until 27/06/2011
DEMEI	Departamento Municipal de Energia de Ijuí	0.36764	29/06/2010 until 28/06/2011
ELEKTRO	Elektro Eletricidade e Serviços S/A.	0.36604	27/08/2010 until 26/08/2011
ELETROPAULO	Eletropaulo Metropolitana Eletricidade de São Paulo	0.29651	04/07/2010 until 03/07/2011
EDEVP	Empresa de Distrib. de Energia Vale Paranapanema	0.33151	10/05/2011 until 09/05/2012
EEB	Empresa Elétrica Bragantina S/A	0.36454	10/05/2011 until 09/05/2012
ENERSUL	Empresa Energética de Mato Grosso do Sul S/A	0.43062	08/04/2011 until 07/04/2012
EFLJC	Empresa Força e Luz João Cesa Ltda	0.39923	30/03/2011 until 29/03/2012
EFLUL	Empresa Força e Luz Urussanga Ltda	0.31736	from 30/03/2011
ELFSM	Empresa Luz e Força Santa Maria S/A.	0.41142	07/02/2011 until

			06/02/2012
EBO	Energisa Borborema ? Distribuidora de Energia S.A.	0.29599	04/02/2011 until 03/02/2012
EMG	Energisa Minas Gerais - Distribuidora de Energia S.A.	0.43907	18/06/2010 until 17/06/2011
ENF	Energisa Nova Friburgo - Distribuidora de Energia S.A.	0.33311	18/06/2010 until 17/06/2011
EPB	Energisa Paraíba - Distribuidora de Energia	0.34886	28/08/2010 until 27/08/2011
ESE	Energisa Sergipe - Distribuidora de Energia S.A.	0.33793	22/04/2011 until 21/04/2012
ESCELSA	Espírito Santo Centrais Elétricas S/A.	0.32889	07/08/2010 until 06/08/2011
FORCEL	Força e Luz Coronel Vivida Ltda	0.36405	26/08/2010 until 25/08/2011
HIDROPAN	Hidroelétrica Panambi S/A.	0.36026	29/06/2010 until 28/06/2011
IENERGIA	Iguaçu Distribuidora de Energia Elétrica Ltda	0.37183	07/08/2010 until 06/08/2011
JARI	Jari Celulose S/A	0.30345	07/08/2010 until 06/08/2011
LIGHT	Light Serviços de Eletricidade S/A.	0.31769	07/11/2010 until 06/11/2011
MUX-Energia	Muxfeldt Marin & Cia. Ltda	0.32609	29/06/2010 until 28/06/2011
RGE	Rio Grande Energia S/A.	0.38429	19/06/2010 until 18/06/2011
UHENPAL	Usina Hidroelétrica Nova Palma Ltda.	0.44479	19/04/2011 until 18/04/2012

i want morebooks!

Buy your books fast and straightforward online - at one of world's fastest growing online book stores! Environmentally sound due to Print-on-Demand technologies.

Buy your books online at
www.get-morebooks.com

Kaufen Sie Ihre Bücher schnell und unkompliziert online – auf einer der am schnellsten wachsenden Buchhandelsplattformen weltweit! Dank Print-On-Demand umwelt- und ressourcenschonend produziert.

Bücher schneller online kaufen
www.morebooks.de

VDM Verlagsservicegesellschaft mbH
Heinrich-Böcking-Str. 6-8   Telefon: +49 681 3720 174   info@vdm-vsg.de
D - 66121 Saarbrücken   Telefax: +49 681 3720 1749   www.vdm-vsg.de

Printed by Books on Demand GmbH, Norderstedt / Germany